Creolia x Six

by

Aneita Lorene

Copyright © 2020 Aneita Lorene

All rights reserved.

ISBN:

DEDICATION

I dedicate this book to the Giffords and the Vinson's. My children and grandchildren and great grandchildren. Special thanks to friends and relatives that encouraged me. Mom and dad, this is dedicated to you in heaven.

I hope this book will make you smile

Chapter one

What just happened?

It seemed like yesterday when I was in Black John Kentucky laying in the grass, singing with the birds swaying with the trees and dandelions tickling my ear, no problems whatsoever! That was a long time ago, so it seemed, but in reality it wasn't that long, it was only 6 years ago! Going to school moving to Cleveland Ohio meeting my friend Ola Mae here and having all the fun that we had together, and living in that beautiful house. My daddy encountered sickness and he overcame it. My Mom and Dad got saved in a church with me and then before I knew it, the blue eyed angel came into my life.

So what do you think I would be doing at this time in my life just graduating from high school and enjoying my beautiful life. I am working in the church with the children on Wednesday nights and having all this time being around the people I love at church and Auza. I don't know what else to say but I am so appreciative of God and how he has taken care of my life and has moved me further in this little bit of time. It's been exciting and it's been enjoyable. Auza and I have always had good conversations, we work together well at the church and we laugh a lot and play. We are getting married on October 16th the day that I will be a wife to this handsome man.

I really don't know what Mama and Daddy are really thinking but I know they could not have picked someone better than this man to be in my life. I will have the exciting life of a Pastor's wife, who else would want to take this place? This is a wonderful place for me and I'm so excited! I do worry about Ola Mae and I worry about Lulabelle, I worry about everybody! Time is passing and I'm going to have a whole church to worry about! I need to get my thoughts right and my ways set to serve the church and my husband. I think that would be the better way for me to be

Mama came in while I was writing in my journal talking to myself as I always do, and she asked me if I could go to the store with her. She wanted to look for a wedding dress for me and of course I'm so happy to do that! Mama and I go to the store and we're looking at wedding gowns and suddenly Ola Mae shows up and lulabelle shows up and suddenly Bertie shows up and Katherine and all my friends at the Christian bookstore show up! Mama threw me a surprise bridal shower, I was a surprise!

So here I am getting ready to get my dress and everybody's here with presents in their hands! Balloons everywhere and it's all in this bridal shop, the cake is on the table and everybody is smiling and so happy, I am beside myself! We had a great shower and we went and ate some cake and drank the juice and had a good time! Everybody's laughing and everybody is all excited about my marriage. Mama gives me a present and it's a blue box that she had for a long time that kept a pearl bracelet in it. She wanted me to wear that at my wedding, because I love pearls!

A nice thing that happened was Ola Mae made me a beautiful ribbon for my hair, this ribbon wasn't a normal ribbon. It was braided with silk strings and it went all the way down my back with cream and pink and white strands. This was so so pretty! My sisters gave me a gift that they all went in together and that was some pretty pajamas and a sexy silky gown. They also gave me an Iron Skillet to knock him in the head when he misbehaves! I laughed so hard at that!

 My boss at the Christian bookstore gave me a beautiful book and it's a book about how a marriage should work and how to get along and to keep peace. I'm excited to read that because I do need to know what I need to do and not to do as a wife! I like to know what he's to do and not to do as a husband. I'm excited about reading that book. It was a beautiful day and it was a beautiful party and I am so glad that we all had a wonderful time together. Mama and I packed up everything and we went home.

As I'm putting things away in my bedroom my mama comes by and says, "Creolia, you may not know this but I couldn't be happier in my life than to have you marry this man, I think it's the best thing that's ever happened to this family," she said with her eyes filled with tears. My heart went out to my mama and I was so happy to hear her feel this way and I know my daddy felt the same way too! In time we'll move on into the future and I will be a wife of a pastor!

As I got ready for bed I was imagining my life with my new husband. I could imagine me going to the kitchen making him coffee, giving him toast, helping him with his studies. To hear what his sermons are going to be for that Sunday, me telling him about the children I'm working with and his opinion on what I should do in certain situations. I was in awe thinking of what a wonderful life this will be with this man. I raised my windows in my bedroom to hear the rustling of the trees, the sound of the locust humming. I am always thinking of Black John Kentucky and always feeling the breeze from the mountains in my face. I will go forward in this big city of Cleveland Ohio and I will do what I can to make it a better place, my thoughts as I drift off to sleep.

As I woke up from my sleep I opened my eyes and thought, what in the world has just happened? They threw me a bridal shower yesterday and I'm actually going to be married in a month and my life will be different from here on! I hear Mama in the kitchen humming and singing. She's making some bacon and gravy and biscuits for my Daddy and me and Lulabelle, I guess I need to get up.

"Mama what was that song that you were humming? I don't think I've ever heard that one before," I said with puzzled eyes. Mama smiled and said, "you've never heard of Swing low sweet Chariot coming to carry me home?" I said "no I never heard that song before but I like it. Can you sing it for me?" Mama said, "how about tonight me and Daddy will play it and sing it tonight for you while we sit around the fireplace?" "Can I invite Auza over also?" "Absolutely you can," mama said, "we would love to have him here we'll all sing some songs!" "Great and I will make that meatloaf dinner tonight!" I love to cook and that is his favorite so far. He really didn't like my Spaghetti and meatballs. I think maybe I had too much tomato sauce in that, I will have to redo that sometimes.

I am getting ready to go to work today and I will not be working after I get married because I will be giving my full-time to the church and to my husband. My boss understands and knows that's the way that Christians are to be so I hope he doesn't get too upset. Auza came by at work and said hi to me. I invited him over to the house tonight for meatloaf dinner and to sing some songs.I would like to share our thoughts with Mom and Dad. He said he would love to do that.

Auza and I sat down and talked at the bookstore and he asked me if he could bring his mother and his two sisters over to the house tonight to meet my parents. I said, "I would be glad to have them all over!" He said, "by the way, my sisters play the mandolin too and they would like to play with your mom." I said, "that would be great!" It should be a happy evening with the whole family together with the mandolins and the songs and the meatloaf dinner. The more the merrier, right?

Mama and daddy really liked his sisters and his mother. They played some folk gospel blues so they called it. Daddy knew a lot. Mama wasn't really familiar with those songs. The madilon is a sound almost like a little guitar. I know it looks like a tiny guitar but they sound like it too."In the sweet by and by, we shall meet on that beautiful shore." What a beautiful song. I notice whenever Eza really gets into a song she looks over yonder like she is singing to the angels. She is a real serious woman.

Mauga is not as serious as Eza, she is a little more outgoing. It's almost like she is the sister that is in control, in a nice way. Auza's mama is so tiny and fragile. She wears these pretty little dresses that flow below her knees. She will sit there and softly clap her and softly tap her feet. I don't see any resemblance of her and her daughters except maybe some facial features. Auza has a good family, and caring sisters.

While I was examining my new inlaws they woke me up from my daze. "Creolia, what song would you like to hear?" looking at me in a real serious look. "Oh I like any songs." I said with a smile. "Well surely you have a favorite" Eza said. "How great thou art?" I said. "Oh yea that is a good one, go ahead and start us out," Eza said. I have never sang in front of anyone like this, Lord help me. I started the song and they all smiled. My daddy got teary eyed. I don't think he has ever heard me sing like that! When I was done they all clapped their hands. Shuwee! that was embarrassing!

They got their things together to leave and Auza's mom kissed me on the cheek. I think that was her seal of approval. Mauga said, "good dinner and thanks for everything. If you don't mind we would like to come back". Wow that was the first! Auza kissed me and gave me that angelic smile. "See you tomorrow," he said. "Yes, tomorrow," I said with dreamy eyes. I am humming and getting ready for bed and Llulabelle walks in and asks me to help her with her homework. As we started on her math, I didn't realize she was a wiz at this. I said, "Ok and why do you need my help?" She said, "Not with math. I need your help with english. I have a spelling bee and I need you to take these words and quiz me on them." I said, "oh, ok, I can do that!" Lulabelle will miss me even though I am up the street.

You know I am noticing something about myself. The palm of my hands sweat when I get nervous. This wedding has me in a head spin! I think I am more nervous being a wife than anything. It is a big deal and I am nervous,excited,anxious and want it to be over.I will pray and ask God to bring peace upon me. I am going to open my bedroom window and feel the breeze. As the breeze passed over me it was like it put a blanket covering me and said goodnight.

Chapter 2

The Great Reunion

I get home from work and I'm really anxious! I am trying to clean up the house, trying to get the meal ready and thinking about all this company coming to celebrate our wedding. Mama walks in and says "Creolia, why are you all over the place?" I said, "well I wanted to make sure the house is pretty because Auza is bringing his sister and his mother over tonight to meet you and Daddy." Mama said, "what?" I said, "I'm sorry but he asked me if he could bring them over to meet you guys and I couldn't say no," I said with a surprised look. "Oh it will be okay your daddy and I are fine with that. The house is clean. I cleaned it up today just go make your meal we'll be fine," said Mama.

I'm making a meatloaf meal and mashed potatoes and green beans. Mama was making the biscuits and her homemade banana pudding. I put on a pot of coffee and the doorbell rings. I run to the door and there he is with his mama and his two sisters. I'm feeling at this moment, way excited! His mama is just a little woman. His two sisters are taller than him so I guess he has his mama looks and his sisters has his daddy, I chuckled to myself. He walks in and his mama is very very sweet and quiet and his sisters walk in with their mandolins. We talked and had a good time and we ate dinner and everybody laughed and enjoyed the moment.

As we get away from the table we go into the living room near the fireplace my mama grabs her mandolin, Auza's sisters grab their mandolin and daddy grabs his harmonica and they sing, Swing low sweet Chariot ... was such a pretty song, I loved every moment of it. Auza and I sat holding each other's hand and cuddling each other on the couch, while the family entertained us with all their beautiful music. It was a wonderful evening to behold!

As everyone was getting ready to leave we all said our goodbyes and Auza he gave me a kiss on the cheek and told me he will see me soon, and then they left. My mama and daddy said they really loved his family and that they were so happy! I can't wait to see him in church Sunday! I then go in and clean up the kitchen dishes and think about all the stuff that's going on and my mind is just rolling in thoughts. When I go to my bedroom I pray and I thank God for the day and I write in my journal these beautiful words.

To behold such a site and to love with all my might. I am so thankful to the Lord, for giving me this day that I adored. As the ribbon that Ola Mae made for my hair and how it intertwined and was so silky and so fine. My beautiful life I could not ask for any more than what I am receiving right from the potters door. The Sky is blue, the clouds are grey but my heart is singing The Lord's praise. Be still my heart for tonight I need sleep be still my mind for tonight in memories, I will keep. As I wait til tomorrow morning to live another day I will be excited as I was yesterday! I close my eyes, I say a prayer, to the Lord, living in his care.

Three weeks and that will happen soon! Auza is picking me up to view the house that we will be living in. What if I don't like it? What if people come over all the time? Oh my God what if the church people don't like me! Am I ready for this? I just want to marry him, I really don't want all of these people at my doorstep! Oh God help me and give me the strength! Suddenly a knock at the door. "I will get it mama it's probably Auza."

"Hi Junior, I haven't seen you in a while. How are you doing?" I asked. "Well things are not great, not great at all!" Junior said with anger in his voice. Junior is my brother and had moved into another house with his wife that he eloped with. "What is going on?" I said with concern. "I would like to talk to daddy," Junior said. "Ok, let me get him!" I hurried to the sitting room where daddy had his morning coffee.

"Daddy, Junior needs to talk to you." "Well tell him to come on in!" He said. "Junior daddy wants you to come inside." I said. "No, I need to talk to daddy out here!" Said Junior with anger. Daddy heard him and went outside. Whatever was said daddy ran inside and got his jacket and told mama he would be back soon. They skid off in the car and I don't know what happened. Oh my Lord what in the world could it be?

From what I heard later on is that Junior got into a fight with his wife and she left him. He wanted to go find her. On their route to go find her she was in a car accident. They are waiting for details. It was at a bridge in the neighborhood. The Policemen were there and people all around. From what I understand the car that she was in flew off the bridge into the water. Was it an accident or suicide? Mama started crying and daddy was shaking. "Junior, come home with us tonight. The policeman said it will take the day tomorrow to pull the car out of the water." Daddy tugged on him.

Junior was in a daze and could not speak. Daddy and mama were supportive. Auza came by and we had to delay seeing the house until we got some answers. Junior began to confess his sins. He said he had been a little rowdy and got angry with his wife and she became scared and ran off. Junior said he is so mad at himself. He hung his head and cried. Auza prayed with him and we all cried and bowed our heads in tears.

The next morning wasn't good. The policeman came by and said that she was in the car and was dead. The policeman wanted to take Junior down to the police station for some questions. Daddy went with him. The whole family came over to console him and give support. It was a grey day and there was no sunshine this day. Juniors wife's family was appalled by what happened, she was their only daughter. They did not talk to Junior because that was not the first time it happened. It was just last week she ran to them because she was scared. Oh, needless to say it was a very sad funeral. Junior left after that and we haven't seen him in days.

Auza came by after the funeral and asked if we could go and look at the house to get away for awhile. When we drove up to the house it was a long brick ranch home. As we walked inside there was a big hard wood floor with a built in fireplace. The house was plenty big! It had 3 bedrooms and two baths. Very cozy and warm. The backyard was fenced and big with a concrete patio and a built-in BBQ! Oh wow, I needed to see something like this after all of the troubles that we went through. I want to move in now and get started with my new life. Auza got ahold of me and swung me around the room and landed a big kiss! We laughed and hugged and started planning our new life.

We drove to our favorite cafe and got our rootbeer float. Life and troubles fall away when I am with him. When we drove up to the house I asked how we would get furniture for the house. He said that the church has a budget already reserved for us. So Auza and I will go this next weekend to get the furniture. Excited! I have never had brand new furniture before!

As I ran into the house Junior was there in the living room talking with daddy. I ran over to him and gave him a big hug and told him he could visit me in my new house anytime! He said he would take me up on that. Lulabelle came running in and said, "me too!" I said "well of course you are welcome anytime!" Mama said, "now, now quit talking like that right now!" Mama didn't like the thought of me being gone. "Lou you will be at my house all the time," I said laughing. Needless to say I will miss them dearly.

Lulubelle came over to me and grabbed my hand and said, "can we go out on the porch and swing?" "Well of course Lou lets go," I said smiling. While we sat there and swung I could hear mama and daddy talking. It was really quiet tonight. "Creolia, can I ask you a question? What does love feel like? Is it what they say? I mean like butterflies and a dizzy head," she giggled. "Lou, it is exactly that and more! I can't get him out of my head!" And yes he makes me dizzy, when he kisses me." I said with all honesty. "Creolia, are you scared about your wedding night?" asked Lou. "YES!" I am very scared," I said with eyes wide open. Lou started to laugh. "Don't talk about it, it makes me nervous!" I said tickling her. "Ha,ha I wouldn't want to be you, that's for sure!" as she ran off the porch. I chased her around the house and caught her and tickled her some more. I sure will miss her a whole lot. We ran into the house and got ready for bed. As I closed my eyes I was dreaming of that day. Lord help me through this I pray. Then I drifted off to sleep.

Chapter 3

The Wedding

Two days and I will be Creolia Vinson! I am beside myself. I am scared in a way. I hope he will like me. I think I need to talk to mama about the events to come. So I go into her bedroom and she is laying down. "Mama can we talk?" Suddenly a knock at the door and it is Katherine and Bertie here to see Junior. Junior was still asleep but I think I will take advantage of this opportunity!

"Girls please come inside by the fireplace." Mama and the girls came in. "Ladies I need your advice. You know I am a virgin and never have done anything. Please let me know what to expect." Katherine and Bertie had a lot to say! I did not know anything and was totally surprised! Mama said "it sounds scary but it will get better." I really don't know what she meant by that until later in life. We laughed and I blushed a lot at all that I heard. Katherine and Bertie are not shy at all!

The day of the wedding was nerve wracking! Ola Mae was beautiful and Lulabelle was gorgeous! The church was full and people were so nice. as I walked down the aisle people were so nice and so cordial it was a wonderful day! As I walked down the aisle Auza's blue eyes were staring right at me and I melted within my own shoes. "To have and to hold until death do us part"..."YES!" We had about 100 people and a table full of food to feed an army! We laughed and intermingled with everyone..until the time to leave. Wow! So quickly and the time has come to consummate our vows.

The next day got here quickly and it was a good morning and it was not that bad (if you know what I mean). I kissed his forehead and I went to make coffee. He was smiling and caressed me in a very sensual way. "Did you enjoy it," he asked. "Yes I did!" I said. He laughed and hugged me and got up for his coffee. "Hey where is that silky gown," he asked. "Oh wow I forgot! I said. Go get it!" He said. I went and put it on and like that we were back in bed! I thought oh this will be fun! What a great day! Wow! What an angel that is all mine!

We spent the whole weekend together and Sunday came with his sermon. Unconditional Love! That was the best sermon ever! He was talking about us! I love you Auza! People were so happy and mama and daddy and Lulabelle were there. I invited mama and daddy over and oh yeah Lulabelle! They came over and we shared stories and songs. Lulabelle will stay with us next weekend. Life is good! Ola Mae came by on a weekday and she was surprised at the beautiful home. "Creolia! You are so blessed! I can't believe that you are married!" Ola Mae's eyes glisten with glee.

We went out on the back patio and had some hot tea and chatted about her new boyfriend. She has a Spanish man from Toledo Ohio. He is 5 years older than her. His name is Danny.
 He moved to Cleveland two years ago. He works with her dad. She really likes him! "So when are you getting married" I asked Ola Mae. "Oh probably soon. I am pregnant, Creolia," she said with shame. "How far along are you?" I asked. "Probably about two months," said Ola Mae. "He has an apartment and we are going in two weeks to the justice of the peace," she said. "Does your mom and dad know?" I asked. "No and they won't know until I get married. Hey can Auza marry us?" Ola Mae asked. "I don't know but I think the proper thing to do is have Danny schedule a time to talk to Auza," I said with a wink. Ola Mae agreed that that would be good.

We had a great chat and Ola Mae needed to get home and talk to Danny. We hugged and she left. Auza comes home and swings me off my feet and holds me all night long. "Have I told you lately that I love you?" I asked. He just smiled and kissed me. We had so much leftover food from the wedding that I really didn't have to cook. I did find out one thing, with his meals add green onions, Green onions and tomatoes! I had my mama's recipes that were given to me as a wedding gift from her. I can't wait to start cooking good meals!

We stayed busy with church, visiting the sick, weddings, baby showers, and church. This is not a boring life at all. I do tell Auza everything at night. I told him about Ola Mae. He agreed that they need to get married. Auza and I are going to visit Junior tomorrow, he is very depressed. We will go and pray with him and comfort him. Sometimes people just need a hand to hold. My hand will be one of those. Junior is still with mama and daddy. He can't bring himself to go home. I don't blame him. In this world we will have our troubles but be of good cheer He has overcome the world! There is nothing too big for God to handle.

It was a week ago we got married! We have had so much fun and stayed busy. I have the house looking good. I need more kitchen appliances, we can go to the Goodwill store this weekend. Oh yeah Lulabelle is coming over. I haven't had a one on one talk with her in awhile. She is a freshman in high school! I am sure she is dying to talk! "Are you ready to go to the Goodwill store?" Auza asked. "Yes, but we need to go and get Lulabelle, I promised her to spend the night. Is that ok?" I asked. "Yes, but I wish it would have been next weekend. Please talk to me first before you invite anyone over," he said. "It's been a busy week and I wanted time with you," he said. I said, "well how about I tell her next weekend," I said. "Ok," he said. You know I like that about him, he wants to be with me!

I went over to the house and Lou was home. Mama was taking a nap and daddy was at work. "Well?" said Lou. "Oh my Lord, Lou it hurt. From what mama said that will go away. I don't know how people do it Lou!" I said with my head down. "Do you have to do it from now on?" she asked. "Yes, Lou, that is the obligation of a wife." I said with dread. "Hey, I came over to let you know we will be busy this weekend. Do you think you can come over next weekend?" I asked. "Sure, anytime is ok with me." she said with a smile. "Ok, I have to get home and make dinner," I hugged her and left.

I got home and Auza was reading the newspaper. "What's new today?" I sat on the arm of the chair looking at the newspaper. "Well did you know they were building this big hotel here in Cleveland?" he looked up at me. "No" looking back. "Look, they are done and it is big! It is Westlake Hotel and this says that it will have a big pool inside. They are doing a special Christmas opening." He showed me the picture. "Wow, that is huge!" I said. "Also the Cleveland Browns will be playing the New York Hankee." Then he turned the page. I kissed him on the forehead and told him I talked to Lou. "Well, what did she say?" Auza asked. "She is perfectly fine with that. Oh bologna, I forgot butter to buy at the store!" Auza put his paper down and said, "What did you say?" "I said I forgot butter at the store." I said with my hands on my hip. "Oh, I thought I heard you say bologna," he laughed. "I did say bologna," I laughed. "That is my slang word when I get frustrated." I laughed. "Oh ok I will remember that next time I want bologna," he laughed.

Oh what a joy this is sitting with my husband. I still can't believe it. We went out on the patio. It was a cool day, you can tell it was going to be winter soon. Our coffee was steaming. We sat there and chatted about this and that. We got up and decided to go to the goodwill store to look for some extra things for the house. As we were strolling the isles we ran smack dab into Mr. Gardner and his wife. "Well well, what a pleasant surprise," he said. "Mr. Gardner, how are you doing?" I asked. "Well it was hard finding your replacement but I think I found the perfect match," he smiled. "Who," I asked. "A beautiful little girl named Lulabelle," he said with a wink. "I just saw her and she never mentioned it," I said, all excited. " I am telling her today in about an hour," he looked at his watch. "That is wonderful, Lou would be perfect!" I said. "Well it was nice seeing you but we need to go, you guys come and visit sometimes, ok?" he asked. "Yes we will," I gave him a hug.

We shopped some more and I was so happy to hear such good news. Auza bought some shoes and I got me a new dress. We went home and I began lunch and I was so happy. "Auza what do you want on your bologna sandwich," I asked with a chuckle. He said "mayonnaise and mustard, and make sure I get a green onion and if you have tomato put that on the side." "Oh bologna, I forgot tomatoes" I said with my hand on my hip. "Ummm, Creolia, we have them in the garden." he said with a chuckle. "Oh my oh my what is wrong with my head, you have my head spinning," I said to him with a wink!

As I was getting ready for bed I was thinking about how Lou would do at the christian bookstore. I know she can greet people and do the register, I just can't see her offering help to someone who is wandering in the store. She just seems like she is in her own world at times. Maybe I will talk to her about it. But I will put it nicely so I don't hurt her feelings, even though it's the truth! "Auza can you bring me some water before bed?" "Yes, I recone I can," he said. "It is getting chilly out, we will need to cuddle so you can keep my feet warm," I said with a giggle. "Don't put your cold feet on me," Auza said. "Maybe I will put some socks on," I said. As we lay in bed Auza said he was thinking about something. He said "You know I think that at church tomorrow we need to schedule a baptism. We can get a vote on where and when before the weather gets too cold. I was thinking about the river maybe next Sunday." "Yes, if it's not too late!" I said.

As we got up we got dressed and ready to go. Auza announced the baptism day and the people agreed. So he said, "next Sunday we will all meet on the bank of the river and bring your baptism clothes. I think there should be about 5 that need to be baptised. We need to get some music together. Maybe Mr. and Mrs. Gifford can bring their music and my sisters bring theirs. Anyone else that has an instrument bring yours too." He smiled at the thought of it.

The next Sunday we all gathered on the bank of the river and handed out the sheet for singing the perfect song:

Shall we gather at the river,
Where bright angels he has brought,
With it's crystal tides forever
Flowing by the throne of God

Yes, we'll gather at the river.
The beautiful, the beautiful, river.
Gather with the saints at the river,
That flows by the throne of God

Ere we reach the shining river
Lay we every burden down,
Praise our spirits will deliver
And provide our robe and crown.

Yes, we'll gather at the river.
The beautiful, the beautiful, river.
Gather with the saints at the river,
That flows by the throne of God

Soon we'll reach the shining river,
Soon our pilgrimage will cease,
Soon our happy hearts will quiver
With the melody of peace.

Yes, we'll gather at the river.

The beautiful, the beautiful, river.

Gather with the saints at the river,

That flows by the throne of God

Mama and daddy got baptised. What a wonderful day on the bank of the river. "Sometimes life just flows like a river"

Chapter 4

A Surprised Honeymoon

Lulabelle was fine about next weekend, she said she did have a lot to tell me. I told her I would see her and mama and daddy at church tomorrow. After church we went to the Goodwill store and Auza found a new guitar! I got some tambourines for my children's choir. I found some iron skillets and some silverware. Oh yeah, two flowered dresses! I love the Goodwill store! We got home and settled in and I was so tired I was out like a light! I awoke to the birds singing and the fireplace burning the next morning. "Auza?" As I looked into the kitchen. "I'm in here Creolia" he was making the coffee. He handed a cup to me and said "taste your coffee," as he sipped. "Too strong!" "Well put some cream in it, it should be perfect" he said. I go and get some milk and yes a perfect cup of coffee.

Sunday morning was so nice to come to church with my better half. All of the people were so sweet and happy about us. I sit in the second pew in my new flowered dress. Everyone's voice sounded so cheerful. Sometimes it seems that heaven has met earth. Before Auza went into his sermon a member of the church said she needed to make an announcement. She got up and said, "the church was thinking that the pastor should have a honeymoon. The church got some money together for you to go somewhere." She called Auza and me up to the pulpit and handed us $200! Auza accepted the check and told the church how much that meant to us. Wowwee, what a blessing! My mind began to race to where would we go? When church was over Auza and I was jumping for glee!

As we were walking home I said, "What a blessing! So where do you want to go?" Auza said "Well I don't know, let me study about that.. Go make some strong coffee with two scoops and we will sit on the back patio and decide." I made the coffee exactly like he wanted and brought it out on the patio. Auza grabbed my hand and said, "Creolia, I know where we can go." I looked into his eyes and thought ..where? He said with a gentle voice and a tear in his eyes, "Black John Kentucky." I was stunned and speechless, "really?" I began to tear up and he hugged me and said, "yes you are going home!" "Are you sure?" I asked with a twinkle in my eyes. "Yes" I would like to see the place you came from. We hugged and the coffee was even tasting better!

Home, hmmm...wow! My mind immediately started imagining Black John Hollow. My house was at the end of the holler. I don't think Auza has ever been in a holler. I don't know why they call it a holler, except you can holler and your voice would carry over the mountains. I would think that old cabin we lived in is still there. People don't change much in the holler. All I care about is sitting on the side of the hill, where I would lay back in the tall grass and gaze into the sky. Nature at its best is right there on Black John

Hollow. At the end of the road not too far away is Ohio River. I can hear daddy out on the porch playing his harmonica and mama rocking in her rocking chair. Us kids playing tag in the front yard. Junior would tackle Sherman and Elvie would come to his rescue. Me and Loulabelle would play in the trees and see how high we could climb. I loved the green apple trees. One day I ate so many that I had a stomach ache all day. I never did that again! Katherine and Bertie would sit on the front porch with mama and daddy.

We would walk down to the river and swing on the grapevines and jump in that cool streaming water. We would grab some crawdads to take home for mama and daddy. To tell the truth, that was also a bath for us. Mama would always say dunk your heads in the water.. When we weren't at the river we were playing in the woods in the holler. Me and my sisters would stay up there in the holler and play house. We would take fallen branches from the trees and make furniture. We would find big leaves and make hula dresses. Always pretending to be a big shot!

My brothers would "try" to build a treehouse. Sometimes the branches wouldn't twist that way. They were really good at building a fort! While us girls would be playing, the boys would hide behind a tree and capture one of us. They were having a fort with prisoners. I would really get mad at them for that! They would laugh and run away, it was funny! We would always find pretty flowers to put in our hair. We would find berries and put on as lipstick. It's funny how you can get into a role playing until you hear your mama calling. You go from being a princess to being a regular kid!

Auza and I could have a picnic right there on the river! We can hold hands and walk up that dirt road and listen to all that nature has to say. I am excited! While I was thinking Auza came into the house and said, "well do you want to leave Thursday and come back Monday?" "Yes" I said with excitement. Thursday morning got here quickly and we began to pack and into the Ford Coup and off we went. We actually had a radio in the car only country music played but I loved Loretta Lynn! I think she lived near Quincy. We had to find a hotel and unpack and get dressed for the occasion.

"So what is the first thing you want to do?" asked Auza. "I would love to go to Johnny's Hideaway, the best fried chicken you have ever tasted!" I said with eyes wide opened. We immediately went there and it was only 3 blocks down the road. Auza had so much to eat that I thought he was going to founder! "Now would be a good time to walk up the dirt road!" I said. "We can walk that fried chicken off!" "Really?" Auza said. "I would prefer to go somewhere and have some coffee." "Oh ok there is a place up the road called Mountain Brew and it is good coffee!" I said with glee. We walked up the road with a struggle of a heavy stomach. I could already smell that coffee a block away. I thought I had died and went to heaven! The coffee is so good, the smell, I remember when mama and daddy would bring their cups full of that aroma. I am going to buy them some and take home for mama to brew. "Well what do you think?" I asked Auza. He likes his coffee black, full throttle was for him and piping hot! "Wow, how do they do that?" he asked. "How can they make such a good cup? I am hooked, we will have our morning coffee right here!" he said.

We walked back to the car and up that lonely dirt road. There weren't any sounds except for the engine of the car. We drove up at the end of the dirt road and over to my right was "the hill" that I layed on to watch for the revenuers. We pulled up and when I got out of the car I thought heaven had just met earth! "Wow, I can smell the honeysuckles, the woods, the green grass, this is the best ever!" I said with a smile. Auza was looking at that little cabin that seven kids lived in. "How did you guys sleep in a little place like this?" he asked. "Well, the boys slept out on the porch, and in the smokehouse on rainy and cold days." We walked into the back where my daddy made his moonshine. Everything was left there as if he just left. I walked into the smokehouse and rodents and cats took over. "Ugh! I hated walking back here in the mornings" I said reminiscing.

We walked into the front yard where I layed and we both sat on the hill and just held hands and listened to nature talk. "Can you believe I laid here on the ground for three days?" I said. " I really can't comprehend a mother doing that to her child." he said with a sad look. "To this day mama has never apologised to me, she has a screw missing I guess." I said shaking my head. "Hey, let's go back to the hotel and rest up." Auza said. I said ok, actually it kind of depressed me. I need to get over that incident of my mama leaving me alone for three days I guess.

We had a wonderful time and went by the river and had bologna sandwiches and pepsi! Such a great time! We drove down to Portsmouth and visited Katherine and Bertie. They both seemed happy. Bertie had a nice house she cooked for us. She can really cook! We went out on their screened in porch and she served us a chocolate pecan pie with a pile of whip cream on top! Chester was a laid back man and always smiling. Bertie was due any day and she was big! "My Bertie when are you due?" I asked. She said "any day now" Is there anything I can do for you?" I asked "Oh no I have Chester family here and Katherine is right up the road, thank you though", she said with a smile. "Yeah she's got me and nothing she can do about it, ha ha!" said Katherine. "You are probably going to get pregnant too, Creolia!" she laughed. "Know telling what may happen, sis," I said with a smile."If I am, I hope it will be a boy!" I laughed. We really enjoyed ourselves and my sisters have not changed a bit. It was good to laugh and play around.Katherine and Bertie will always be special to me. I am so glad God helped them to find a good life. I waved goodbye and we all threw kisses. My eyes watered as I said goodbye. Auza was quiet; he knew being here brought back so many memories. There were bad ones and there were good ones. All in all it was a nice and peaceful feeling.

Chapter 5

Family and Friends

I think I am more in love with Auza than what I was! I can't believe as crazy as I was for him that it could be possible to love him even more! Not just physical but spiritual. He knows a lot more than me when it comes to the bible. I will make a New Years resolution to read the whole bible through. At least I can understand the what, where when and how about the bible! There is a huge difference that we had about our home life. I was raised with cursing, drinking and rowdy parents. He had a very christian mother and sisters. I understand people more on the street life, so maybe he can use my help in that area. I have substituted my own curse words. I say bologna when something is not working, I say foot when I stumble or hurt myself, and My Lord when I am surprised. I guess everybody has their vise, I chuckled to myself.

When we got home the first thing was to do laundry and unpack. Auza had to check on members and make sure all was well. He left to do that. I love being a wife and having a church. I get to try out one of mama's recipes. I think I will make her sausage and cabbage. Oh bologna, I am out of green onion! I guess I can walk to the store. It is only two blocks. I will go ahead and see what else I need for the week. Auza gave me $30 of the money the church gave us. As I was walking Ola Mae and her boyfriend pulled up and asked me for a ride. I said sure! When I hoped in, we talked about Black John and she wanted to know every detail. They came into the store with me. Ola Mae said that Danny wanted to talk to Auza tonight if that's ok. I said sure just come home with me and he will be home soon.

We got home and I put everything away and made some sweet tea. We sat outside on the patio. I asked Danny about his home life and his family. He said his family worked a lot and the kids had to help at times. He said that there were five of them two boys and three girls. He was the third child. I asked if he is ready for marriage and he said I was born ready! We all laughed. As we were talking Auza walked in. I introduced him to Danny and he sat down with us. Danny started to talk to us about marrying Ola Mae. It was very pleasant. Auza asked them, "when do you want to do it?" They said, "this weekend. We are tired of sinning." "Ok how about Saturday around 2:00 here at church?" asked Auza. "We wanted it to be very casual," said Danny. "We are doing this on our own and telling the parents later." I had to give my opinion on that.

"Now Ola Mae this only happens once I feel you need to ask your parents to the wedding." Ola Mae looked at Danny and asked him. "It will be ok with me, it wouldn't hurt to ask." Danny said. "Ok 2:00 at the church," said Auza. "Creolia, do we have enough dinner to share?" asked Auza. "We sure do just settle back while I make it," I said with a big smile.

What a blessed evening that was! We were like old friends. I didn't realize what a sense of humor Auza had! I saw a different side of him! Now I can tease him without him getting offended. I love a good laugh! We ate some chili and cornbread and Danny had two of everything! Ola Mae said it was the best! We hugged and said goodbye. After they left Auza said he went by mama and daddy's to check on Junior. He said that Junior was doing better. I haven't heard from Sherman or Elvie lately. I hope they come by sometimes and say hi. Sherman and Elvie are married but working at the same factory. I finished the dishes and we went to bed. We were both tired. Don't get me wrong we were still on our honeymoon!

We were both early risers around 6 or 7 when we got up. I made some coffee and you could feel the wintery air hitting Cleveland Ohio. Oh my Lord I am going to have to get into all of my winter clothes. "Auza, you know it is going to be cold soon, do you have any winter clothes?" I asked. "Yes I do but my clothes are at mom's house," he said. "How about we go there today and get your things?" I said. "Sure how about after lunch, I have to study for my sermon tonight," he said. "Yes I have to get my songs together for the kids," I said. I guess we have got a pattern going on how things work. I love working with my kids at church. They are much different than the kids I hung out with! They love their little jingle lings to play.

When we were done we drove to his mama's house. They have a two story house with a big front porch. His sisters worked at a factory together. They seemed very close and secretive. I could hear them whispering in another room.

They finally came out and said that they were wondering if we would like to come over on Sunday for dinner. Auza immediately said "Yes!" I thought they must cook well, his eyes were wide open! "Creolia, what kind of food do you like?" as they stared at me with a very serious look. "Ummm, I like anything," I said with a smile. "Creolia, do you cook for Auza?" asked Muga. "Yes I do" I said. "Creolia, do you wash his clothes and iron them?" asked Eza. "Yes" I said. "Creolia, do you know when his birthday is?" asked Eza. "Well I am not sure," I said. They gasped and looked at each other like I committed a crime! "Auza's birthday is June 17th," they said.

"Creolia, are your parents saved?" Muga asked. "Yes they got saved at the church under Auza's preaching," I said proudly. "Creolia, what does your daddy do?" asked his mama. "Well in Black John he made moonshine and sold bottles of whiskey to the neighbors, but now he works at a factory." I said with a smile. "Does your daddy still drink liquor?" asked Eza with a frown. "No, he replaced the bottle with a bible," I said smiling. They then relaxed a little bit for now. My Lord, the look on their face was as if I just said that daddy murdered someone, that look I will never forget, silentently I giggled.

Oh boy that was rough I felt like I was being drilled! How in the world did Auza live with that! As we drove home I was quiet and so was Auza. I think they embarrassed him. His mama was very quiet. She didn't even resemble them at all! "Auza, do your sisters have boyfriends?" I asked. "No, they do everything together, believe it or not they are still virgins." said Auza. "I don't ever see them getting married," he laughed. "Don't go barefooted around them, that's a sin," he laughed again. "How in the world can being barefooted be a sin?" I asked. He said, "they said it is showing your nakedness!" he chuckled. "Will they be asking me more questions?" I asked. "Probably!" he said. "And they will remember every answer too, believe me!" he said with all seriousness. "My Lord you don't have a chance with them," I said. "The only thing that saved me was my mother!" He said. "Well I am in shock no one in my family would do that" I said with a sigh.. "Creolia, you have to understand, they have nothing going on in their life at all. I guess you can call them busy bodies. " "Your mom isn't like that at all," I said. "Yeah, mama is the sweetest woman on earth," he said with a sigh.

The next day I got tickled about Eza and Mauga. When I saw the look in their eyes as I told them that my daddy made moonshine, that was hilarious! I won't tell Auza I will keep my chuckles to myself. You know there are moments that make you laugh and this was one! Auza walked in while I was having a chuckle. "What are you giggling about?" asked Auza. "Oh something Ola Mae said to

me that gave me a little snicker," Oh my God I just lied to my husband! I will have to ask God's forgiveness! "Hey Auza, what should I take Sunday?" "Oh make that peach cobbler and we'll drop by the store and get some vanilla ice cream," he said calmly. "Oh ok!" I can make a good peach cobbler. It is Wednesday and we have youth service tonight. The kids will have tambourines and bells. I love singing with them. Their little feet march around the church, while they play this song;
"I may never march in the infantry or ride in the cavalry or shoot the artillery. I may never fly over the enemy, but I'm in the Lord's army. I'm in the Lord's army, I'm in the Lord's army. I may never march in the infantry but I'm in the Lord's army, Amen!"

It is Saturday and Ola Mae and Danny are going to be here at 2:00. I went to Woolworths yesterday and got them two throw pillows with their initials on them. One with an O and one with a D. That will be for their bed or couch. I will make some punch and her mama will bring a cake. I guess her parents took it well. What can they say? She's pregnant! Me and Auza ran over to the church to make sure everything was good! When they got here she was beautiful! She had a baby blue dress and a baby blue veil. He had a dark blue suit and a baby blue tie. His parents got there first and we chatted with them and then Ola Mae's parent's got there and they introduced them. It was very strange because they just all found out that they are going to be grandparents! I can't imagine how it would be if they hadn't told them at all! The right thing was done.

"Do you Ola Mae take this man as your husband to have and to hold until death do you part?"asks Auza. "I do," she said with tears in her eyes. "Do you Danny take Ola Mae to have and to hold until death do you part?"asks Auza. "I do," said Danny. "I introduce to you Mr. and Mrs. Gonzales!" said Auza with a big smile! We all clapped our hands. His parents were so happy, not so much for Ola Mae's parents. I went over and hugged her and told her that I loved her.The reception was nice and it looked like the parents all got along! Ola Mae hung close to us and I can tell she was very relieved and happy that they had told their family. "Ola Mae", can you believe this? Here we are in Cleveland Ohio and both married!" I said with tears in my eyes. "I know it is crazy Creolia!" she said with glee. "So how many children do y'all want?" I asked. "Well if you had a choice?" I laughed. "But if I did have a choice I would like 3!" Ola Mae said. "How many do you want?" "If I had a choice maybe 5?" I said with a giggle. We had a lot of fun and exchanged addresses in case we wanted to send each other a card. Auza invited her and his parents to church. The cake was delicious. It was marble on the inside and blue and white on the outside. Her mom put a man and women on top that looked like them. His parents brought jalapeno poppers and salsa dip. His mom made some spanish roses on the pews with ribbon flowing down the sides.

We drinked some punch and I think that was my first time having poppers and salsa. I think I will make some salsa. I walked over to his mother and asked her for the recipe. She gave it to me: 3-½ cups of tomatoes (about 4 large tomatoes) 1 large green pepper diced. 1 medium onion diced. 1 serrano pepper diced and seeded, 1 jalapeno diced and seeded, 1 whole lime squeezed, 1 Tbsp of sugar, 2 tsp of salt,1 clove of garlic finely chopped.

Auza loves spicy foods. He even likes pickled pigs feet! I can't wait to try this on him, he will want it all the time! I think all foods are good but southern food is the best! As we said our goodbyes we all hugged and exchanged addresses.

"Hey Auza what is your favorite food," I asked as we walked home from the church. "Your meatloaf by far." he said. "What I am asking is what interesting different kinds of food do you like? I only cook country food and I would like to try something different." I said. "Well I had what they called pepper steak in the Three Seas. Everything was sliced in strips. You cut the steak long wise in strips and green pepper and roll out some dough and cut it in strips. You put everything in a skillet and put some oil for it to cook and soy sauce. You cover it up until the dough strips are done." he said. "Yum, that sounds good, I am going to make that for you!" I smiled as we walked into the house.

So much to learn about each other! This can go on for years! I don't know but my life has changed so much that I can't believe it is me. Who would have known that I would marry a preacher man. I am happy and willing to do anything to keep our life happy and easy. Auza goes and starts studying for his sermon tomorrow and I think I will try out that pepper steak for dinner. I get our clothes all ironed out and ready for tomorrow. Winter is coming and I will do some canning. My mama taught us girls how to can anything. I would like to make some peaches for jam, some apples for apple butter, some black berries. The vegetables would be canned green beans, tomatoes, cucumbers pickled, That should keep me busy this fall

Chapter 6

The Choice

Wow how time flies by. It has been 3 months and guess what? I am pregnant! I just have a little bump but the church people are acting like a president is going to be born! They want to throw me a shower I guess that will happen soon. They are the sweetest people in the world! Ola Mae will have her baby 3 months before mine. Our house will have a nursery. Goodwill will have plenty of things for me to purchase. A couple of the ladies at church gave me maternity dresses. They are so cute! They have green flowers on one (my favorite color!). The other one has blue ribbons on it, it's a skirt and a top.

The ladies threw me a shower and it was so funny. They had the funniest games. The prizes were just as funny as the games. I got a lot of blues from them. Do they think we are having a boy? They had a punch that had floating pacifiers They had a cake that looked like a rattle. The room had ribbons and baby things all over. I think I got enough stuff for four kids! They had a naming game that if it's a boy's favorite name and if it's a girl's favorite name. We ended up with David and Sheila. I don't think that will work for me. I thanked the ladies and Auza helped me with packing everything up. They went way beyond on this!

I will have to train someone for my position on Wednesday nights in a couple of months. I wonder what I will have? We will find out on July 2nd! It was Sunday night and every Sunday night we have confess your sins one to another. It is kind of fun because it makes you really aware of your actions all week. And when you goof up you say I will tell that at church Sunday night. Auza's mom and two sisters came to church that night, not aware that we have that every Sunday night. So my little white lie that I told Auza, my "Chuckle" was about Ola Mae, when it was actually the look in Mauga and Eza's eyes when I told them my daddy made moonshine. I will say I told a little white lie, as my confession. Hopefully Auza won't ask me about that.

When all started confessing Eza and Mauga began to whisper really loud. I heard Eza say, "I didn't sin any" and Mauga said, "I did, I got mad at a co-worker." So when it came time for Eza to stand and confess she blurted out, "I gossiped about someone," then quickly sat down. To make a long story short the next Sunday night Eza was the first to stand and she said, "I want to confess that last Sunday night I lied about sinning. I didn't gossip about anyone, I just didn't know what to say!" she sat down quickly. Oh my, I thought I was going to die with laughter, I had to run to the bathroom to let it out! I laughed so hard that my eyes watered. So she sinned "in her confession of sinning!" Oh my Lord, that was funny!

I don't know but I need to journal some funny things that happen in church. This happened last Wednesday and Sister Cox always stands up and says "Praise his holy sweet name, I can just see the angels sitting up there on the rafters." Well over the weekend we had the ceiling fixed..no more rafters. So she stands up and says "I can just see the angels upon...what happened to the rafters?" she said. The whole church burst out in a loud laughter. She laughed too, that was fun! I don't know if it is because of pregnancy, but I giggle a lot! Everything seems funny! I guess the joy of the Lord is my strength.

Another funny thing happened at church. We were sitting in church and kept hearing this knocking. It was like a hammer knocking lightly on the floor. When we all sing the knocking would get louder. When everything calmed down it would stop. This was really driving me crazy! I asked the lady in front of me, "did you hear a knocking." She laughed and said "Oh that is sister Beard, when the spirit hits her, her feet start shaking, that is her hills knocking on the floor." she laughed "Watch out though if she really gets the spirit she will run!" We both giggled.

The time had come for me to have my baby. We went to the hospital and the baby wouldn't come out. 24 hours past. The doctor finally took Auza behind closed doors. We have some terrible news. The baby is breached and if we force the baby out your wife could lose her life. My question for you is which one do you want to save? Your baby or your wife? Auza immediately said, "my wife, save my wife!" Auza went to church and started a prayer chain. All of the people were praying for me. Somehow they must have prayed through! Because the baby was saved and me! Auza was so happy about the life of our little boy being saved! We were so close to death and God spared me and the baby. Auza said he wanted to name our son after his father, Albert Gamale. So our beautiful little boy was named Albert. Even though I survived I needed a lot of help. I had a lot of stitches and I had to be careful.

Everyday was a process but eventually I became better. I laid there and had magazines to look through. I really am no good when I have to stay in bed. "Auza, could you go to goodwill and get me some yarn and needles. I would like to make some booties for Albert." I asked. "I was going that way anyway, I need some tools. What color of yarn?" he asked." I would like Blue and white," I said. "Ok, I will be back, is there anything else you need?" "No, just make sure it is crochet needles," I said. "How did you learn to Crochet?" He asked. "My grandma taught me and Bertie" I said. I loved my grandma, she was always happy. She had so much talent! She could sew, and cook, and play the piano. No wonder my daddy a;ways smiled, he had her to raise him.

Auza got home and gave me a basket full of yarn and needles. "Wow, thank you!" I said. "Creolia, they had what you call a basket fillup sale for $2!" "I would have got two," I laughed. "Well now you can make all you want," he kissed my forehead. I was anxious to get started and this will keep my mind occupied. Booties, sweaters, a little hat for his head, I am so excited!

Auza was getting his sermon ready for Sunday and he came in and asked me. "You are new with the bible and the bible stories, right?" he asked. "Right," I said. "Ok, I am going to ask you some basic questions that a babe in Christ should know, Ok?" "Ok", I said with a little resistance. "Why is it important for a newborn in Christ to be baptised?" He asked. "Symbolic that my sins have been washed away?" I said. "You have that partly right, what other reason could there be?" he stared at me. "Because the bible says so?" I said with my thinking cap on! "No, to show the public your outward commitment to Christ," he said. "Ok, one more question. "What is the purpose of communion?" he asked. "To show that we drink his blood as symbolic of his blood that was shed for me. The bread is symbolic of his body that was sacrificed for me," I said with a smile. "Correct!" he smiled. "I like that, that made me think," I said laughing.

"Hey Auza are you using that in your sermon?" I asked. "Yes I am but you will be surprised what the subject matter will be. I am preaching on "Symbolic witness." Things we do that should show we are christians," he said with pride. "I love that! That is wisdom from God," I said with a wink. "I was thinking, what do we do so that people would know that we are christians," he said as he walked back to studying. "Just telling someone that you will pray for them, having compassion and giving to the needy. Holding someone's hands during hard times." I yelled while he walked away. He walked back and stuck his head around the door, "Sinners do that, that is why we show commitment to Christ, through baptism and communion." He winked at me.

You know sometimes he seems a little cocky. I am going to read through the whole bible and I will be like him! It is really strange that my mom and dad never taught me anything about Jesus. They would tell me the Christmas story but no more than that! I would have thought my daddy was a christian because he loved me like no one ever did until I met Auza. I am so glad he gave his heart to God., him and mom! Albert is being such a good baby. His eyes stare at me while he nurses. I start humming a tune and he falls asleep with me while we take our nap.

How do I describe how I feel being a mom? I feel blessed to have a baby so innocent and trusting to cuddle up in my arms. Albert will always be special to me no matter what, he's a survivor! The journey through our life is but a blink of an eye. Yet God allows us to enjoy every single minute. Holding a baby that comes from Auza is a miracle for me! Just knowing this child has me and Auza in him. Wow, what a miracle to behold!

Chapter 7

Red Roses for a Blue Baby

Auza came to the door and said "Mauga and Eza are going to come here and cook meals and take you to the bathroom. I am going to the store. Is there anything you need?" "Yes a pepsi and some dill pickles and potato chips." I said. He chuckled and said "You sure you don't want icecream on top of that?" "Yeah and some butter pecan ice cream please!" I don't know how I can be in this condition and want all of that! God is making me better!

Auza left and Mauga and Eza came running in with a pan of hot sudsy water and washed me. Wow, that felt so good to get all of that sweat off of me. They gave Albert to me to hold in my arms. He has all the milk that he wants! Oh my I hurt so bad, I hope this pain will go away. Auza brought some pain medicine home and I have to take it every 3 hours a day. I fall asleep and wake up to Eza and Mauga playing their mandolin's. "There ain't no grave going to hold my body down" I like that song. My mom and dad came over and they did not know that I was bed ridden. So Auza told them of the battle we had. I talked with them and daddy's eyes filled up with water. I said "Don't cry daddy, daddy please don't cry. I am better! Both me and the baby!" He laid his head in my lap and hugged my belly. Mama stood there rubbing my head.

Eza comes in and says, "we have to take her temperature." When she did I have a slight temperature and that was not good. Eza told my mama and daddy, "She needs to rest, the doctor said if she gets a temperature to get cold towels!" They ran into the kitchen and wringed out cold towels. Oh my Lord please help me! "Auza go and get some ice cream, that cold ice cream could help," yelled Mauga. Auza ran in there and brought it in and daddy fed it to me. "That's my baby girl, you will be alright I promise," he said with tears in his eyes. Auza said, "I want us all to do something. Everybody gather your hands together and we are going to pray for healing!" Suddenly a harmony of angelic voices began to pray, praying for me! I got chills all around me and felt the presence of the Lord. They prayed for at least a half hour! I then fell asleep and they quietly left the room. I then went into a dream..

There was an angel that was standing at the end of my bed. The angel raised his two wings and began fanning me. He said, "Be still and know that I am God" He then took a vial of ointment and poured it on my head. When he did, it felt like it flowed through my whole body. I saw a cloud come up from my body and it left. When I awoke my heart beat slow and in the night I saw a glow. This was no dream he turned my way and then I heard him say. "You have been healed and you will live. Go and tell it everywhere you go. For you my daughter has found favor!"

The next morning I yelled, "I am healed! Eza, Mauga, Auza!" I screamed. They came running in. "I am healed, take my temperature! Hurry so you will know!" I said with anticipation. Auza took it and wala 98.5! "Ladies and Gentlemen your prayers brought heaven to earth! I had an angel that visited me last night and he fanned me with his feathers and poured ointment on my head. He said you have been healed and you will live. Go and tell all for I have found favor!" I said with eyes wide open. Mauga and Eza began praising God and rejoicing. I was tired of lying in that bed! I got up and made some coffee and rejoiced all day long! God is an awesome God!

The next couple of days was the best. I think I ate that whole jar of pickles. Well I think I left Auza one or two. We went to church and the church people rejoiced and praised God. I guess you could call it a jubilee! Boy oh boy I am feeling stuffed, blessed, and happy! This is the day that the Lord has made and I will rejoice and be glad in it! Daddy got some red roses from my daddy! He said "here baby girl, some red roses for a blue baby" with a wink. What beautiful flowers. After I got them I didn't wan to throw them away and I wanted to preserve them. I found out you can't do that. So what I did was I bought me a rose bush at the store. I grew a big rose bush, enough to keep my table pretty all summer.

We named our boy Albert and he had his daddy's blue eyes.If I could describe that sweet little fella in one word it would be lover. He loved kisses, cuddlies, and food. He was such a joy to have. We dedicated him in church because that's what christians do and he wasn't supposed to be here according to the doctor. God is truly good! He started walking at 10 months old and loved to run. I was out in the backyard and he found a turtle. He picked him up and put him in his little pocket. I said "Albert you are going to smother that turtle," I said laughing. He didn't know what that meant so he brought it over and took it out and petted its back like a dog. I went and got a pan and put some water in it and some greens from my garden. I think we kept it for a month, until one day we left the pot outside and the next day it was gone. Albert cried his eyes out. My heart ached because that was his first heartache.

We had been at this church for a year and a half. The church was really getting big and the responsibility was becoming overwhelming. While we were having a bible study a man had mentioned that he had a friend at this church that needed a pastor. The church was in a country setting with the Blueridge mountains surrounding it. When I heard that my mouth dropped open. It's in Marion NC and right now they have about 50 people at the most. Auza was ready for a new venture. So we prayed about it and God sent a pastor where we were. Before we knew it we were moving! Albert helped Auza with small things. A man at church helped us with 2 pick up trucks. The new church had been so long without a pastor they were excited! All eager, I found out some information about this place.

Marion is a city in McDowell County, North Carolina, United States. It is the county seat of McDowell County. Founded in

1844, the city was named in honor of Brigadier General Francis Marion, the American Revolutionary War Hero whose talent in guerrilla warfare earned him the name "Swamp Fox". Marion's Main Street Historic District is listed on the National Register of Historic Places. The population is 4,038.

This is the photo the man at church gave me, and all I can see that big mountain around it.

The good thing is it is only 117 miles from Cleveland. We will be there in about 2 hours. Albert was a trooper; he loved trucks! He would sit on Auza's lap and help him drive. We were following the trucks and driving in our car. The church people packed us a big lunch with potato salad and my favorite bologna and tomato sandwich. They even got Auza's green onion and pepsi cola! I can't wait to be back in the mountains. As we arrived at our new house it was a block from the church. A yellow framed house with a little front porch with two rocking chairs. There was a man sitting on the front porch from the church. We got out of the car and he ran up to us and shook our hands.

"Welcome to Marion! We have been so excited to have you here! We've been without a pastor for almost a year! It has been hard." he smiled. Auza said, "well we are glad to be here and help out." "Come on in here and look at your new home." He said. "Oh by the way my name is Earnest," he said with a smile. We painted the whole inside and put in a new refrigerator. It's not well isolated so in the winter you will want to wear extra clothing. You have a fenced yard with a place for a garden. The last pastor grew some delicious tomatoes and cucumbers. The store is two blocks away." He then walked into the bathroom. "We cleaned everything so that you don't get someone else's germs.We all bought you groceries." He began to open the kitchen cabinets. "We got you dairy products so you can cook."

He finally stopped taking and Albert said "ruck"looking at the stranger. Albert ran to the door and said "ruck". Auza started laughing "He wants us to unload the truck!" I laughed and Earnest did too. We motioned for our truck drivers to come inside. Albert started running all over the house. That was his way of saying he approved the house. We got all settled in and sat out on the front porch, people would wave as they passed by. In small towns everyone in town knew what was going on. I stayed busy with the new house, missing my mama and daddy though. I can hear the locust up there on the mountain. At night I can identify every bug I heard. They say there is a lake not too far up the road. Auza loves to fish and I do too. We will catch some fish to fry. Albert will love catching a fish but he probably would want to pet it. I chuckled. To have the hills right up under my nose, I so bad want to go and roam the woods.

"Hey Auza, why don't we have a picnic up in the hills Saturday?" "No not this weekend Creolia, I have two sermons to preach Sunday. You know on Saturdays I like to prepare myself. We have a long time here so wait, please!" he said firmly. Well I guess the hills can wait! "Ok but you know the hills are calling" I said with a chuckle. Auza had bought some seeds to plant. Green onions two packs, tomatoes three packs, cucumbers two packs,corn seeds, and Okra and green beans. This will be a lot of hoeing. I went and got a hoe and gave Albert a shovel. He would dig a hole and play with the worms! Well, it looks like we will have no problem having worms for fishing. It got too hot to finish. I went and made some sweet tea and relaxed while Albert played.

As I sat on the back porch I could view those beautiful mountains surrounding us. I closed my eyes and I breathed in the fresh air that smells like fresh jasmine. While Albert played I went into a daydream. I was feeling the wind in my face and I heard daddy say. "Baby girl come sit on papa's lap." he pats his leg. I ran onto the porch and jumped in his lap like I had done many times. He said "Let's sing our favorite song, ok?" "Yes papa" I said. "You are my sunshine, my only sunshine, you make me happy when skies are grey. You'll never know dear how much I love you so please don't take my sunshine away." When we sang that song everything in this world was perfect to me. "Mama, mama!" said Albert. Albert nudged me out of my daydream." Look" he said. I jumped out of my chair. He had a worm with a thousand legs. I said "Albert put that down!" He put his hand on the porch and laughed. Talking about a heart attack he made my heart pop out!

Come here Albert I have a song to teach you. And I taught him "You are my sunshine," we sang it three times. Now he will run into my lap and say, "mama, let's sing sunshine," and we will. I chased him around the house and as we were coming around the corner we ran into Auza and he caught Albert and swung him around and around. Albert giggled . "Auza can it be any place prettier than this?" I asked with wonder. He laughed and said, "it would be better with some food to eat" he laughed. "Well tell me kind sir what would you like to eat?" I laughed. "How about some fried bologna?" he smiled. "Oh, that will be easy to do," I tucked my arm in his.

Chapter 8

Come On People Now, Let's get Together

I am nervous about meeting strangers. I hope they like me and I like them! We get out of our car and the people are nothing like Earnest. They barely said hi. As I go to the second pew everyone else sits behind me. Oh Lord you are going to have to help me on this one, I mumbled to myself. The Sunday School superintendent got up and introduced Auza and me. The congregation clapped their hands. They have a song leader that gets up and leads everyone in songs. They made some announcements and then gave the service back to the superintendent. Then they had several people to do a special song. Then they gave the service over to Auza. He talked a little about us and where we came from. His first sermon was on "symbolic witness", which was a very good sermon!

After church everyone was talking to each other but not to us. I have never seen such odd behavior in my life! As Auza and I walked home Auza said, "I got a really odd feeling, those people seem really distant!" He chuckled. I said "I think we will have to work on their hearts. I would assume they have been hurt by previous pastors. It must have been really bad hurt, because once someone has experienced a bad hurt it takes a lot of love to get them back." "Good observation Creolia" Auza seemed enlightened. We got home and I made a pot of coffee and started lunch. Auza went into his office I guess to pray and get in touch with God! Albert ran around like no care in the world. Albert will be a year old tomorrow! I will have to see what would be fun for him. I did see a park down the road, maybe take him there and have a picnic. As I was cooking we got a knock at the door. Auza went to answer it.

"Hi Rev.Vinson my name is Willard. Umm, I have a problem and I need your advice." "Sure come on in! Would you like a cup of coffee or some tea?" "No maybe later," he said with his head down. "Well come into my office and we can talk. Creolia I will be in the office with Willard," Auza said. "Ok," I said.. "Well Albert I guess it is just you and I," and then another knock at the door. I ran to the door and a wide eyed lady said, "Is Willard here?" I said, "yes but he is talking to the pastor". "Well he is lying about everything and I don't want a tainted name! Let me in there too!" she said. "Well, wait right here and let me go and tell him.Auza there is a lady out here that wants to be in on the conversation." "No!" said Willard. "Creolia, can you take her out on the front porch and hear what is on her mind?" asked Auza. "Well ok," I motioned her to the front porch.

"Can I have your name?" I asked. "Yeah sure my name is Lisa and I am Willards wife. I am very upset!" she started crying. "I worked down at the Food Mart grocery store and Willard came by my work and couldn't find me. I was actually in the break room having my break. My boss walked in and said something really sweet to me and I got up to hug him, that's when Willard walked in. He slammed the door and walked out, I ran after him and he just ignored me!" she began crying.

Needless to say I really didn't know what to say, thought Creolia. "Lisa, have you and him had problems in this area before?" asked Creolia. "Well yes, after work my boss took me to the coffee shop just to relax from a hectic day and Ms. Nosey rosey saw us and told Willard!" she rolls her eyes. "Can I ask you a question?" asked Creolia. "Sure" she said. "If you heard that Willard went to a place with his female boss and you found him hugging her, what would you think?" "Well, I would not like that at all!" she said. "Well then maybe you should be a little more sympathetic in his conduct. If he didn't love you then he wouldn't care." Creolia said. "I think you need to reassure him nothing is going on, and I would stay at least 3 feet from your boss at all times," I chuckled. "Yes, you are right." She said.

Auza came out with Willard and Lisa ran to him and hugged him crying saying that she is sorry. They left arm in arm. "Wow, I am glad we got that resolved," Creolia said. "Whatever you said to her worked," Auza said. I don't know but I think there is more to this church than what the eyes can see. We get ready to go to church and right out on the front lawn two women are arguing about a cat! To beat all a cat? Auza went up to the ladies and the one said that her cat was stolen. She said she saw it on the other ladies front porch. The other lady said that cat has always been hers. Auza said go get the cat. They both lived in walking distance. The lady went and got the cat. Auza said you two ladies stand 10 feet apart and move 10 feet away from me. Now give me the cat. The lady handed him the cat. He pet it to calm it down. Now both call his name. Just one time. Auza let the cat go and it went straight to one of the ladies. He said "I reckon that cat is hers. How long have you had the cat?" Auza asked. She said, "5 years" He asked the other lady and she said "2 months." So there you are, the wisdom of king Solomon.

We settled in at church and Willard and Lisa walked in, hand in hand and sat right next to me. Just when you wouldn't think things could get any better Auza preaches about the wisdom of King Solomon. You know sometimes life just flows! I really love this man and I could sit here all day and listen to him preach. Tonight a few people came up to us and introduced themselves. Albert wanted to hug everybody. God sure gave me a bundle of joy. Earnest bought Albert a ball and bat for his first birthday, even though I don't think he is old enough for that yet. As we walked home I asked Auza if going to the park for his birthday was good, he said that's fine.

We went to the park for Alberts birthday and he loves the slide! He wore me out! I made us a nice lunch and made cupcakes with chocolate icing. Albert ate two of them, I laughed so hard he had chocolate everywhere even on his eyes, that was so funny! Auza cleaned him up while I gathered things together. Time passes so quickly. While we were getting into the car Auza asked, "would you like to go for a drive up the mountain?" I said "YES!"

We drove right into heaven, we had our windows down. I stuck my head out the window and got a sniff of the tall oak trees. Albert put his head out too. His cheeks jiggled from the wind, his blond hair was going every which way. As we got to the top of the mountain there was a view that would take your breath away! We sat up there talking and laughing, oh my Lord it was so much fun! I didn't want it to end. Albert found a big red maple leaf and brought it to me. I will keep this forever. Auza picked a handful of daisies and gave them to me. Life is good, as I sat there feeling blessed.

"Auza, if you had to do it all over again what would you have done differently?" I asked. "Hmmm, I think I would have gone to some kind of bible school. That way my knowledge of the bible would be better." he said as he looked over the mountain chewing on a piece of grass. "How about you," he asked. "I would have not gone to school that day when mama left me. That way I would have gone with them to Cleveland," I said looking down with my finger playing with the grass.

It was nice talking about things and having an open relationship. I looked over and Albert had a piece of grass in his mouth like his daddy, I laughed! Auza came and sat down and said, "Can I tell you something that I have never told anyone?" I said "well of course you can." "Remember when I told you that I hopped trains and went here and there? Well there was this one train trip that there were several hobo's in the same cart. When I hopped on they started pushing me around and the train was going really fast. I couldn't jump out so I fought back. In the middle of wrestling we both fell out of the train and we went tumbling down the hill. I got all scratched up and he did too. But guess where we landed?' "Ummm, in the river?" I said. "Nope, we landed at the bottom of the hill in strawberry patch. We were so hungry we ate almost that whole patch!" He laughed. We were bleeding and all scratched up but we didn't care we had strawberries to eat!" he laughed. "Did you ever see him again?" I asked. "As a matter of a fact we joined the three C's together. We became cooks for the men." he laughed. "Oh, so you can cook!" I laughed.

We packed our things up and gathered some nature to take with us. I found Auza and Albert a long straw that they could chew on. As we rode down the hill I saw a black bear and said, "look Albert, there's a black bear!" as I pointed. "Oh, look it has babies!" I smiled. We all stopped the car to view the scene and Albert waved goodbye to them. "I am going to tell you a bedtime story about three bears," I said. We rode home singing "You are my sunshine."

Chapter 9

Surprise!

It's the middle of September and our anniversary is in a month. I don't know but I think I would like to go to the top of the mountain and spend the day there. The fall leaves will be on the ground. The cool breeze and the smell of fall. I have a big surprise for him. I told him what I would like to do and he said fine. He asked if I could find someone to watch Albert so we can cuddle. Oh yeah I think that will be perfect. As I anticipate our big date I am in search of a babysitter.

We went to church Wednesday and a young teenage girl seemed to be a good babysitter. I approached her. "Hi, how are you tonight?" "Fine." she looked puzzled. I had Albert in my arms. "I am looking for a babysitter for Saturday. Do you know anyone that would like a little money?" "I would!" she said with excitement. "What is your name?" I asked. "Melisa," she said. Well would this Saturday around 11:00 work for you? We will be gone most of the day. We will probably be back around 5:00." I said. "Sure, that would be fine!" she said with glee. "Have you ever babysitted before?" I asked. "Oh yeah many times." she said. "Good we will see you Saturday."

Saturday came and we went to the top of the mountain. I brought a nice blanket and a basket of food. We sat there with a pepsi cola in hand and just enjoying the moment. "Auza I have a surprise," I said with a wink. He looked at me with a "what in the world" look. "I am pregnant!" I smiled. He grabbed me and kissed me. "Another baby? That is heaven sent!" he said. "I thought with that fever you had, you would not be able to get pregnant." "Well I am a month and a half!" I said with a big smile. We rolled on the blanket giggling and kissing it was the best ever anniversary!

The day brought such joy and the leaves were all gold, orange and red. We went and got a basket full of fall leaves to take home. What a day of remembrance. When we got home Albert came running and Melisa said he was adorable. We paid her and she left. Albert and I went through the leaves and picked out the prettiest ones and pinned them on the walls. Thanksgiving is right around the corner. I think I will make a pot of chili. That would be a good October dinner. I will make my grocery list for the ingredients.

2lbs of hamburger,½ lb spiced sausage, a large green pepper, a large onion, two large tomatoes,one medium tomato paste, a can of original rotel,Chili powder,Paprika, salt and pepper, one garlic clove, spiced mustard,one bag of kidney beans and crackers,½ stick of butter.

I like to cook my beans for about two hours.I keep the soup from the beans which is the water I would have added. I like to roast my green peppers and tomatoes and jalapeno on a broil in the oven. Once they are burnt on all sides I deseed them and sliced them and put them in the mix. I cook the hamburger in the spices. I then add all ingredients and put on medium heat for two hours. I like to put in a stick of butter at the very end, that gives it a velvet look and a sweetness to the chili. Auza will love this! Oh yeah crackers on the side!

While I made my grocery list guess who pulled up to our house? My best friend Ola Mae and Danny! "Ola Mae!" I ran out to her. What a precious face to see! "OMG! Your little boy is adorable!" I said with my hands over my mouth. "Creolia, Danny asked me where do I want to go, and the first thing that popped into my head was Creolia's!" Ola Mae said. "The last time I saw you was right before your fever, they wouldn't allow me in to see you. I kept in touch with your mom. I was having little Danny at the same time. The next thing I heard you moved here! I love it here Creolia it is so pretty!" Ola Mae said. Albert and little Danny played with each other so well!. "Come on in! Auza is at the church cleaning and preparing for his sermon Sunday. I was getting my grocery list together, I am making a pot of chili. I hope you are staying for a couple of days!" "Oh, yes if you don't mind." said Ola Mae.

"Hey, do you want to take me to the store so I can get the chili started? We can go by the church and let Auza know." Creolia asked. "Sure that would be fine." said Danny. We got to the church and Auza was out pulling weeds around the church. "Hey Auza look who came by for a visit!" Creolia said. "Well well, good to see you!" Auza said. "We are going to the store because I am making a pot of chili. Is there anything you need?" Creolia asked. "Pepsi, milk, green onion, bread." he said. "Nice seeing you I will see you later at home." He smiled. "Nice seeing you too!" said Ola Mae and Danny.

You could hear my laughter clear down the street I would think. We had so much to catch up on and stories about our kids. Two peas in a pod Ola Mae and me. The chili was smelling so good. "Creolia, can you teach Ola Mae how to cook?" asked Danny. "Ha! Ola Mae I would love to write out some recipes that you might like." Creolia said. "Danny, loves chili, beans and potatoes, good fried chicken, gravy and biscuits, pinto beans, do you know how to make that?" Ola Mae asked. "Ola Mae I know how to make all of that!" I said laughing. "I will write my recipes down for you. I am making gravy and biscuits in the morning. Come and watch me. We will have fried chicken tomorrow too! Everything starts with an iron skillet, and always bacon at hand. Bacon grease makes a great sauce for anything!" "Thank you, Creolia." said Ola Mae. "Yeah thanks," said Danny.

Auza came home and we all sat down for the chili. "Danny I put jalapeno in the chili but I didn't put in the seeds because that would be too hot for the kids." Creolia said. "Oh do you have any hot sauce?" Danny asked. "Yes I do, Auza likes hot sauce too" She smiled. I think the guys had two bowls and a pack of crackers. "Creolia, I have to say that is the best chili I have ever had!" Danny said rubbing his belly. "Oh thank you!" Creolia said with a smile. "It is a little chilly tonight, Auza can you get the fireplace going?" asked Creolia. "Ok, give me a minute to digest, can you get me a cup of coffee?" Auza asked. "Ok, Ola Mae, Danny do you want some coffee?" "I would," said Danny. "Not me I want some milk if you don't mind," said Ola Mae. "Ola Mae are you pregnant?" asked Creolia. "How did you know?" asked Ola Mae. "Ha! Because you want milk! I am pregnant also!" said Creolia. Oh my Lord we are going to have babies together again! When are you due?" asked Ola Mae. "The end of April," said Creolia. "I am too!" said Ola Mae. We both jumped up and down like little girls.

Oh my, I did not want her to leave! "Ola Mae you and Danny are welcomed here anytime you like!" I said as they were leaving. "Hey Creolia, will you send me a picture of your new baby, and I will send you one of mine!" She smiled. "Ola Mae there is the best farmers mart right up the road on your right." I pointed. "Oh ok we will go by and check it out, thanks for writing those recipes down!" "Yeah thank you!" said Danny. "Ha!" I laughed. I threw them kisses.

"Auza they are so cute!" I said. "Hey Auza are you going to be busy this evening?" I asked. "No." he said. "Well, me and Albert started hoeing the garden and it got too hot, do you think you can help me finish?" "Creolia! You are pregnant you should not be out in this hot sun hoeing!" he said loudly. "Ok that is why I am asking now," I said rolling my eyes. "Come on, let's do this and get them growing!" he said. We hoed and hoed and hoed. We will have so many vegetables to eat. I love cooking fresh vegetables. Corn on the cob is one of my favorites! As we were hoeing I would find worms and Albert would come and rescue them. It will be different when we have our next baby. I stopped hoeing and I told Auza I have to rest and he said, "go rest your feet and your belly," he laughed.

As we retired we were both exhausted. I didn't tell Albert about the three bears, I was way too tired. As we layed in bed I felt so refreshed and complete. Only God can do that. "Auza do you feel complete and satisfied?" I asked. "Yes, I feel like heaven has met earth up there on that hill." he said while he yawned. "I hope we can go back soon," I said while snuggling up to him.

Chapter 10

My bundle of joy is pink

I said laughingThe holidays went well. Between the Christmas plays and Christmas caroling and Christmas banquet, they wore me out! I am ready for 1951! My new year resolution is to spend more time outside. I have three months before my next bundle of joy gets here! Auza loves reading the newspaper, "Listen to this Creolia." said Auza. "According to the newspaper they are going to broadcast a colour tv from The Empire State Building! They also have a new cartoon called Dennis the Menace. Huh, colour tv my oh my what is this world coming to?" Auza said, shaking his head. "Dennis the menace what's that about?" Creolia asked. "Oh it seems a little boy is a brat to a guy named Mr. Wilson." We laughed. Ha! Now how come they think that is going to be funny?"

January is a long winter month. I think the hills keep the wind away. We burn the wood in the fireplace a lot. We have been so busy with the church that I would like to have one day of peace. These church people need to consult the lord, they act like Auza is a priest. "Auza, I know what your next sermon should be," I yelled from the kitchen. "Oh yeah, what?" asked Auza. "Seek ye first the kingdom of God! Maybe the people would start going to Jesus with their concerns than with Father Vinson!" Creolia chuckled. "Ain't that the truth!" Auza laughed. "I figured out why they had no pastor for a year. The news got out that this church is very needy!" Creolia chuckled.

The months pass fast and the time has come for our new bundle of joy! The Easter lilies are in bloom and the rabbits are mating. Albert loves it when he sees a bunny. He says bunny raaabit. He is so cute. I wonder how he will be with his new brother or sister? Auza and I are sitting in the living room eating popcorn. Suddenly one sharp pain and then another. "Oh Oh I think it is time, Auza, get my suitcase, it's time" "Are you sure?" he asked. "Why of course my water just broke," I said calmly. He ran around and got Albert ready and off we go! As we were leaving Earnest had pulled up and asked if there was anything he could do. All Auza said is "please let the church people know." Earnest said "sure!"

Well five hours later a beautiful little girl! "Oh Auza have you ever seen such a precious baby in your whole life? Look at those big eyes and her fluffy cheeks!" my eyes watered. Auza had Albert to tend to. He was half way listening. "Well I need to get home and put Albert to bed. Are you ok with that?" "Well, ok but try and get that babysitter tomorrow so that you can spend some time with me...please!" "Ok I will," he said. I can sit here all day and just stare at her. "Rock a bye baby in a tree top, when the wind blows the cradle will rock. When the bough breaks the cradle will fall, down will go baby cradle and all."

Three days in the hospital and now I am going home. On our way home I asked "Auza, what are we going to name her?" "Well I thought her first name should be Robbie, what do you think her middle name should be? Robbie May? Robbie Jean? Robble Lou?" asked Auza "Robbie Darlene" Creolia said. "Robbie Darlene, Robbie Darlene, I like that!" he laughed. We got home and the ladies of the church had food that they brought over including desert! Albert ran up with banana pudding all over his face. "Somebody get me a towel" Melisa came running with a towel. I wiped his face and he smiled. "Albert come here and look at this! When he looked at the baby he kissed her forehead. "Auza did you see that?" "Yes, he is a cutie," Auza said. The ladies bought us a bassinet for our baby. Tomorrow it is going to be like a train station, people running in and out!

A couple of months has passed by and Alberts second birthday! My mama and daddy and Lulabelle are coming down. I am so excited! Robbie is alert and giggles a lot. Albert would go to her and tap his little finger on her mouth and she would giggle each time. We have laughed at him many times, he is so funny! Suddenly a knock at the door. "Hi mama and daddy! Lulabelle look at you! You are all grown up," my eyes watered. Albert came running and gave my daddy a big hug. Daddy grabbed me and gave me a big hug also. "Lou are you 16 years old now?" I knew the answer, just starting conversation. "Yes, I am," she smiled with those big dimples. "Lou you are so pretty, I bet every boy in your school is after you!!" I meant that. She smiled like saying, yep! She has natural curly hair. Big dimples and beautiful green eyes.

Albert was so excited he was running all over the place so Lulabelle took him outside. "Mama you want to see your grandbaby?" I motioned her over to the bassinet. She came over and held her heart, daddy came over and said "Mama have you ever seen such a beauty?" Robbie's big blue eyes staring up at them. Mama picked her up. We all had such a good time. It was as if time stood still. I made everybody some coffee and Lou and I had a bottle of pepsi cola. We sat on the front porch, and played with Albert and swung the baby. "Lulabelle do you have a boyfriend?" I asked. "No not yet, I had one but he was too possessive! I don't care right now." she said. "Well you have plenty of time. Go get a job and occupy your time and make money at the same time!" I said with a wink. "I miss you," she said looking sad. I ran over to her and hugged her. "Tomorrow we are going to the coffee cafe and do some bonding. We may have to take Albert." "Ok" she said with a smile.

"Mama how are the boys and have you heard from Katherine or Birdie?" I asked. "Well Sherman is doing well and he and his wife seem really happy. They have a second child on its way. Elvie is ok I guess I hardly hear from him. Junior has a new woman and they attend your old church! Katherine is fine and Bertie is doing great! I told them about you and they were so happy about all of your ventures. "Well I would like to make daddy his favorite dinner.

Chicken and dumplings!" I smiled at daddy. "Woowee that sounds like a party to me," he did a little Irish jig and made us all laugh! Oh my what can I say, them visiting us made my year! We had such a good visit with mama and daddy! I even had time to talk to Lulubelle and her boyfriend.

You know it is so hard to say goodbye to my daddy. It breaks my heart every time! "Daddy you keep on smiling and make sure Lulabelle gets plenty of attention!" I said as they went to the car. "Mama, you are looking good , your cheeks are so rosey," I said as I kissed her cheeks. "Lou you come her and give me one more hug," I said with teary eyes. "Here daddy some fresh vegetables from our garden," I handed him the bag. As they drove out I stayed until the car disappeared. I went in and cried for a minute. Time passes and I am pregnant again and I am going to have our third baby in April...again!

Time passes and it's mother's day at the church and the church got me some new dresses and hair ornaments. We had cake and punch after church, coconut cream cake with a pineapple spiced punch, which was delicious. We all laughed like we were family. It is amazing how different they are now to us, compared to the way they were when we came. I made a roast in the oven for our Sunday dinner and it really smells good! I am going to get some fresh green beans from our garden. I will make some homemade cornbread. I don't know this for a fact but I think Auza loves my cooking, I know daddy loved my chicken and dumplings! Albert is picky. He will be 2 in July and he is a spark plug! He gets a little jealous about Darlene. She gets more attention because she is a baby. She is 3 months old. I take the kids out on the front porch and we swing on the swing and I sing them a song…

"Doctor doctor where are you come and help get Albert well, he is sick and about to die and that will make his sissy cry. Sissy sissy don't you cry Albert will get better by and by. Dressed in pink and trimmed in blue that's the sign she will marry you. Dressed in blue and trimmed in green that's the sign he will call you queen." Albert clapped his little hands and said, "another mama another." He really loved to sing. I sang a couple more songs then went back to cooking. Robbie is such a good baby. She stares at me when I sing, so cute! As I cooked I noticed Auza. Auza is in there reading his newspaper and if there is anything that makes him excited is cars and patten leather shoes. "Creolia, come here and look at this!" He said his eyes wide open.. "That's a new nash coming out at Christmas!" he said. "Well we will have to get a bigger church if you want something like that! I snickered. "Don't underestimate our God," he snickered. "Okay we will pray for a miracle and maybe the car will roll itself right up to our house!" I laughed.

WE all ate dinner and went out on the front porch and drank our coffee, and as sure as we were sitting there, a brand new Nash pulled up to the stop sign. Auza's mouth fell to the floor. I was shocked to see such a site. Auza says, "I wonder who owns that?" "I don't think it is anyone in our neighborhood!" I laughed. "Man, that was beautiful, I like the blue better than the green," he said. "Thank you God for that!" I said as I raised my hands.

Yours for a Merry, Merry Christmas with the World's Most Modern Car!

We had such fun just watching the neighborhood and the cars and the bikes. Albert loved every truck that went by. "Mama, ruck!" he would say. Someday I hope we can let him ride in a truck. Maybe when fall comes they will have a hay ride.

We are going to see a site that is to behold. Everyone at church says "Have you seen Chimney Rock yet? We are so tired of saying "no" that we have decided to go. This is the article we found at the library.

In 1902, with the financial backing of his brothers, Morse paid Freeman $5,000 for 64 acres of Chimney Rock Mountain, including the Chimney and cliffs. Many small tracts purchased over the years expanded the Park to nearly 1000 acres. Want a more personal experience? Join us for an Animal Encounters program. Meet the animals and hear how our ambassadors came to live with us and more about the habitat, adaptations and natural history of their kind. These interactive programs are fun for the whole family and are offered at 2 pm every day in April, June-August, holiday and October weekends and Christmas week. Check out our events calendar for more details.

That lake there is Lake Lure this is what the library said about it: Lake Lure covers approximately 720 acres (2.9 km^2) and has a shoreline of approximately 27 miles (43 km). The dam's power plant began operations in 1928 with the sale of electricity under a 10-year contract to Blue Ridge Power Co., a local predecessor of Duke Power. I can't wait to go!

Chapter 11

And The Beat Goes On!

We celebrated Abert's 2nd birthday upon the mountain. I am so glad his birthday is in the summer. Even though the temperature is in the 80's, up here on the mountaintop it is 70's. Auza picked me some daisies and we found some ginger roots and sassafras roots. I will save these for hot tea! I also found some mustard greens and collards. I got a whole bunch, I am going to make a batch! The blue berries will be growing in September. We had Albert's favorite cupcakes. This time he ate 2 cupcakes. Robbie she giggles at everything! Auza will make a funny face and she giggles. Albert goes around acting like a clown, she laughs, it is so cute.

We went to Chimney Rock but we didn't go to the top. We sat by Lake Lure and enjoyed this beautiful view. Can you believe that God made this just because. I am in heaven and can breathe that fresh green air. They had a playground and that had red cedar chips. There was a huge slide, I went down it with Albert. There were big monkey bars, Albert climbed the whole thing. They had a merry-go-round that Auza and Albert went on. The lake was so pretty, people were floating down with their ores in their hands. We had our bologna sandwiches and potato salad and of course a bottle of pepsi. A beautiful day and now we can say "yes" we have seen Chimney Rock!

Time passes by and we are nearing September and Auza said "I'm going to the store and is there anything that you want?". Umm... pickles,pepsi,ice cream...please!" "Creolia, are you pregnant? "Again?" staring at me. "Yes I am but I didn't get there by myself!" I laughed. "How far along are you?" "2 months," I said. "Well I guess we will do it again!" Auza said. The only thing that I don't like about having babies is washing the diapers in the bathtub. Actually I wash all of our clothes by hand and hang them outside on the clothesline. The diapers are gross sometimes. Ironically I will have this baby in April! Two babies in April! As Auza gets home he said, "we will use the same bassinet as we did with Robbie."

Time passes quickly and we have our third child. Welcome to the world Merele Edwin! A boy was good! I love Albert, Merele will be good too! Robbie, but I like to call her Darlene, she is a year old and still on the bottle. Shuwee! I have my hands full! Robbie is trying to walk. Oh my Lord, he gave me good babies. Now it takes us longer to get ready and time to do anything! The very next April I had another baby, three Aprils in a row. The third child born in April was Zelma and she was ready to come into the world because she came quickly! Big blue eyes, spry!
It came time that another small church needed Auza's help and that was in Georgetown Ohio. So the church was not happy to see

us go. It seemed once the church became too large then our time was done to go and help another church to keep its doors opened. Georgetown Ohio was only 36 miles from Cincinnati Ohio. It was a tiny house. This is going to be rough. We have four children and another one on the way. I am tired and can't keep up. This church is so little that I doubt if they can support us.

We moved in and these people weren't as country as in Marion NC. They lived near the big city of Cincinnati. Auza saved a little bit of money in NC and there was this grocery store that was up for sale. Just a little bit of money and just take it over. This was God sent! At least we had groceries! Vinson's Grocery. I had our fifth child Aneita. Because my hands were full Auza had to take Aneita in a basket to the store with him. He would prop her up on the countertop and people talked to her all day long. Then I had another child. Our last one was Glenda. Her and Aneita were only a year apart. Robble and Albert helped me with her. Aneita and Glenda were very close. This is what the newspaper said about Georgetown Ohio:

Georgetown, Ohio
Large city - Southwestern Ohio along the Ohio River and Kentucky/Indiana borders. September, June and May are the most pleasant months in Georgetown, while January and February are the least comfortable. President Ulysses S. Grant grew up and attended grade school in Georgetown. The school house is still in the town, where the locals refer to it as the Grant Schoolhouse.

Also in Georgetown are Grant's childhood home and the tannery that his father owned across the street. All three are stops on the Land of Grant tour. Being the county seat, Georgetown hosts the large Brown County Fair and parade which brings in thousands of patrons from other towns. The headstone of General Thomas L. Hamer is in the old cemetery located by the Brown County FairGrounds. Seventeen acres of the city was listed on the National Register of Historic Places in the Georgetown Historic District.[15]What Bert Has To Say About Cincinnati Metro Area Cincinnati is a livable city at the crossroads—literally and figuratively—of north and south, east and west, and Old World and New World. Called by some the "northernmost southern city," it is a transportation and cultural gateway between the industrial North and rural South dating back to Underground Railroad days. The area's largest industry and employer is Procter & Gamble, with a history that dates back to the city's early stockyards when soap was made from animal byproducts. Other companies make soap and cosmetic products, while machine tools are another important industry. The area has experienced growth in financial services and in commercial and manufacturing facilities for overseas companies. There are some businesses supporting the auto industry, but the area's economy has been less susceptible to disruptions from that industry, and is in good shape for a Midwestern city.

We stayed in Georgetown for 4 years. As time went on Auza sold the grocery store and we moved into a farmhouse, Lindsey Grove is a small town near Lawrenceburg Tennessee. This was going to be an adventure for six kids. I actually feel like God gave me a break. The kids had so much to do that they would only come inside for a drink of water. We had an outside toilet so that was convenient! We were excited about living in Tennessee! Another new adventure!

Tennessee

The area of Tennessee was originally part of North Carolina. North Carolina ceded the area of Tennessee in 1790 to the United States, and this area was organized as the "territory . . . south of the Ohio River" on April 12, 1790. Tennessee was admitted to the Union on June 1, 1796, as the 16th state. Upon resolution of a boundary dispute with Kentucky in 1820, Tennessee assumed generally the same boundary as the present state.

Census data for Tennessee are available separately beginning with the 1790 census. The 1790 population shown for Tennessee is for the Southwest Territory, which generally had the present state boundary. Data for the legally established state of Tennessee are available beginning with the 1800 census.

Chapter 12

Out on The Farm

We moved into this farm house with farm animals in 1959. Ok here is the breakdown of my children.

Albert was 10 years old
Robbie was 8 years old
Merele was 7 years old
Zelma was 6 years old
Aneita was 4 years old
Glenda was 3 year old.

Here we are on a farm with guineas, chickens, ducks, dogs and pigs. This was a perfect thing for the kids. Albert and Merele would play all day running through the woods and running down the long dirt road. One day all of the kids went to the store to get some milk. The store was at the end of the long dirt road. As they were walking back something silver was shining in the road. All of the kids ran toward it and it was a silver dollar! They thought they had found gold. They ran home to show Auza what they had found and Auza kept it. But he did go to the store and bought all of us a candy bar and pepsi.

There was something very strange about this house. Cabinets would be opened the next morning. We heard dishes crash on the floors in the middle of the night, and nothing would be there. You could hear people whispering and laughing quietly. Auza would get up and say "Who is there?" One night Aneita woke up crying because someone was shaking the bed. She peed the bed. I hadn't done laundry yet so I put on her a pair of Merele's paints.

Oh my Lord she ran around showing off and acting like a clown! Everybody laughing at her didn't encourage her to stop. We only had an outside toilet and there was just a tiny light. So any time the kids had to go to the bathroom they woke us up. I got in the habit of making them go to the bathroom before bed. I know her accident was from being scared.

Grandma Vinson and Eza and Mauga were coming to visit. Eza said that she felt a bad spirit in the house. She got out her virgin olive oil and anointed all of the door frames. They cried in Jesus' name to leave this house! From then on the spirit was gone. Grandma loved the chickens and they enjoyed themselves with all of the farm animals. Zelma and Robbie went next door to the neighbors house to get some milk. As they were walking home they were both carrying bottle milk and they were swinging the milk. Their bottle of milk crashed together and the glass flew everywhere and sliced Robbie's ankle bone very badly. I had to bandage it up. Here came Eza and her anointing oil and prayed for her ankle. I think she went through a whole bottle of it before they I left.

I think Zelma is one that loves adventure! She found the first arrowhead under the porch. She had a dime in her hand and it fell through the cracks of the porch. Zelma went under that porch and in hunt for her dime she found an arrowhead! After that Auza sent out the whole possy to find more. That gave us a bag full of arrowheads. So you could see Zelma's mind working and she said, "daddy what can I do to earn money?" and Auza said "how about you go out there and keep the bugs off of the peanut plants. So that is how she lost her dime.

It was a very hot day and what I do is go get a big bucket and fill it up with cold water. The kids would strip down to their panties and get cooled off. The dog was jumping in and even the ducks jumped in. We had so much fun watching them all have a cool day on a hot one. Auza came home one day and said, "family I have an announcement. Come in here and listen, this is very important. There is a cotton patch down the road and they need help. They will give us 50 cents for each bag we bring in. What I want to do is give each of you 7 cents a week for doing it. Each Sunday you can go to the store and get you a candy bar and a coke!

Ok and guess who has to make lunch for 8 people every day? Me! This is temporary maybe a month? It was hard for me to see my kids sweat and drag that cotton bag behind them! I brought plenty of water. When we were done at the end of the day they would weigh our bags. The weight should be equal to 50 cents. Well my son Merele decided to ease himself of too much labor so he put a rock in there. The weight was more than 50 cents. They looked into the bag and got the rock out. If they had poured that cotton out to go through the cotton mill it would have broken the whole machine. Merele got a spanking for that one. I wasn't happy that we had to labor like this. What is a family of eight to do?

My poor kids were tired and worn out and I wasn't feeling well either. That hot sun all day long and them bending over and putting that cotton into the canvas sacks. Lunch was a blessing but hard on me. We are out in the fields picking cotton all day and then coming home dead tired. I have to make our lunch again for the next day. We only have one more week to go and I will be happy. I have to say we did it together and I really don't remember the kids complaining except being hot. Zelma would sing for us while we picked. Darlene would make sure Aneita and Glenda behaved. Merele and Albert would throw cotton balls at each other. Auza stayed ahead of the pack.

The last day of cotton picking Auza went and bought everyone a candy bar and a pepsi. The kids were so happy. We made a toast to the last day and said goodbye to the old cotton fields.

Chapter 13

Rheumatic Fever

After cotton picking time was over something strange started happening. I became extremely fatigued, my chest hurt, my legs ached. Auza had to take me to the doctor. The doctor checked and took my temperature and said we need to call your husband in. Auza came in and this is what the doctor said. "Your wife has rheumatic fever. This is very, very serious. Rheumatic fever can last from 6 weeks to more than 6 months. Your long-term health depends on how your heart has been affected by the disease. Rheumatic fever can weaken the heart muscle and affect your heart's ability to pump. The heart valves may also be affected. One or more valves may become scarred and after a while may have trouble opening and closing properly. Damage to the valves may not show up until years after the illness. Eventually, the valve may need to be repaired or replaced with surgery. Starting antibiotic treatment early when you have rheumatic fever may prevent permanent damage to the heart.

It's very important to prevent recurrences of rheumatic fever because the severity of heart trouble is related to the number of attacks of rheumatic fever. You may need to take penicillin regularly for months or years to keep from getting more strep infections. How long you will need to take preventive penicillin (or a different antibiotic if you are allergic to penicillin) depends on many factors. The recommendations change every few years as more is learned about how to prevent and treat the complications of rheumatic fever. Oh my Lord what is happening? Bed ridden, again I had it bad and I was bed ridden for over three months. What a mess 6 children and a husband that does not know how to cook. Auza called the church together and told them that we needed help. So the ladies of the church took turns helping out. Auza had another grocery store and that helped with food. What the women did was really weird. They would go and get towels which were soaked in hot water. They would then take and wrap them around my legs. Then they would take bricks and heat them up and put them at the bottom of my feet. This was a remedy that works with Rheumatic Fever.

Robbie came into my room every single day and asked me if there was anything that she could do for me. I always had something that needed to be done. Glenda was only 3 years old, the baby. She would crawl up in my arms and take her nap. Albert wanted to be "the man" of the house. He would make sure that the house stayed straightened up. He knew I kept a tidy house. Merele stayed busy outside taking care of the animals. Zelma would help Robbie do things for me. Aneita kept Glenda busy and they played. They would come into the house and ask for old pot pie pans. Auza would bring them home for a quick meal. Glenda and Aneita would go outside and get a little water and stir mud into the pans. They would take rocks and put them into the mud for nuts.

They would take green leaves and crumble them up into the pie. Then they would take the mud pie and let it sit in the sun to dry. When the mud pie was done they would bring it in to show me what a pretty mud pie they had made. Aneita and Glenda were always content to play well together. The same thing with Darlene and Zelma. Merle and Albert a different story. They always competed in everything they did. If we could eat the mud pies we would have no problems. One day they made 10!

Talking about pies, the ladies of the church made homemade rhubarb pies. Auza loves rhubarb! They brought 3 pies over. I would sleep a lot, there was a lot of pain in my legs. I couldn't wait for the ladies to come and nurse my legs. I wish I knew why me? Here I am a mother of 6 and can't even take care of them. I know I shouldn't feel guilty but I do! Father God almighty please help me to get better! I started crying seeing how hard Auza was working and the kids all pitched in to help. Robbie saw me crying and she came and laid her head on my hands. Zelma would go and get a warm wash cloth and put it on my feet.

Albert would read me some stories from the books from school. Robbie took Zelma with her to school because Zelma begged to go. The school was ok with that every once in a while. Merele was in the first grade. Three of the kids were at school during the day. Three were home. It really eased the burden when they went to school. Auza would always bring me something from the grocery store. I love my family! Sometimes it was so quiet that it was eerie! Auza went to the grocery store, three of the kids went to school, three girls here making mud pies and finding arrows or playing with the animals. I had no energy for anything.

As time went on I became better after three months laying in bed. I appreciated more of what I have, I loved more, I gave more. Being in bed for 3 months gives you a lot of time to think. My kids really came through and I saw what big hearts they all had. I just want to cook and feed my family! Auza said amen to that! The women of the church I so appreciated. As soon as I get my strength back I will crochet them some scarfs.

When the kids came home from school, the teachers had the students to make me a get well card. I laughed at how cute and creative their little minds worked. I wrote them a thank you note. Auza got me the paper from the store. You know people come through for you when bad times happen. My new year resolutions will be to go that extra mile for people.Try to give the spirit of love. The first thing Auza wanted me to cook was fried chicken,green beans and my potato salad. When I set the table every single eye was on the prize,my fried chicken. There wasn't one leg left and every belly was sticking out! One thing my kids never missed out on a meal. I have so many greens outside that I can cook. I fry up some bacon and put the greens in there, I cover them and cook them until they are done. My favorite snack is cabbage. I love to munch on cabbage. I can make a great sausage and cabbage! I guess I am ready to live again.

The first thing I wanted to do was walk down that dirt road. Aneita grabbed one hand and Glenda held the other. As we were walking I was very weak but it made me strong. The girls talked to me the whole time. Mama this mama that, as soon as Glenda stopped Aneita started. When you have been in bed for months it is different how you see the world. We walked over into the woods and I found a mess of greens. Glenda and Aneita picked every pretty flower they could. I said, "pick a bunch and we will make a pretty decoration for the table.Get some purple, pinks, blues, and yellow." I was pointing to the ones I saw. I got me some good greens and I am going to cook them right now! Glenda and Aneita put the pretty flowers on the table and Zelma and Darlene got the silverware out and the plates. We had some ham from the freezer and some sweet potato and butter. Oh the smell that I have missed. Auza came home and was so happy that I was cooking. As we dressed the table we all were singing "This is the way you dress a table, dress a table, dress a table, all the way home" A happy home it was.

I think the best thing I liked about the farmhouse was that it seemed the kids stayed busy with nature and animals. I would go out to the henhouse and gather eggs for breakfast. I love chickens and roosters. I love to hear the cock a doodle do every morning. We have the prettiest rooster! We really need a cow out here as much as we use the milk. We have a pig but we can't bring it upon ourselves to slaughter him. I had a rocking chair out on the front porch and the kids would play Red Rover three against three. I taught them how to play the game of kicking the can. Then at night I would save my pickle jars so they can catch lightning bugs. One night one of the kids left the lid off of their jar and we all layed there entertained in the pitch darkness with lightning bugs.

We have a dog named Sporty and he is so funny. I think he is the smartest dog that I have ever seen! Albert will go outside and Sporty is his pal. Albert would get a ball and throw it as hard as he could down the dirt road. Sporty would run and get that ball, even if it was off the road. He would bring it back to Albert. He winds at night when the kids have to go to bed. I made him a bed on the front porch with a blanket.. He would chase wolves and foxes away that tried to steal our chickens. He was a great watch dog. All the kids loved him. Sporty was a sport, so the kids said. The best thing was that he would walk the kids to the store and wait for them until they came out. Love our Sporty that's for sure!

With my illness and all the business around here gives me little time with Auza. So sometimes we retire early just to have time together. One thing that he really hates is when I get a craving to eat cabbage at night. I don't know why but I crave cabbage and ice. No, I am not pregnant, that is just me. Well we had a great conversation last night about this church. He said he didn't feel it here it was way too hard on him. That was the first time I heard him make a confession like that. He said he had a meeting with the deacons and they said that we were too much of a burden to the church! Well not a burden just they couldn't afford to support us, which made them feel bad. "So what are we going to do Auza?"I was confused. He said, "I contacted a few people and found out they needed help in Lawrenceburg about 30 miles from here.They said it was a dying church that needs to be saved. It would be hard but not as hard here." he put his hand on his forehead. "Oh it will be ok, we will get through this Auza." I said rubbing his back.

Chapter 14

Back to the Appalachian Mountains

We finally moved out of the farmhouse and have taken a small church in Lawrenceburg Tennessee. I am so excited it is on top of a mountain and in the Appalachian Mountains. I read this article;

The city of Lawrenceburg has a total area of 12.6 square miles (33 km^2). It is the largest city on the state line between Chattanooga and Memphis. Located on the southern Highland Rim, Lawrence County and Lawrenceburg are set atop of a large mountain plateau of the Appalachian Mountain range with elevations ranging between 810 feet (250 m) to over 1,120 feet (340 m). Map of the Appalachian Mountain Range.

I feel so bad for the kids, even though they hated the outside toilet, they loved the animals and country setting. It was a nice big farm house. I guess this is what the Lord has done. We find the strength and the ability to move. After having Rheumatic Fever I am very weak. Robbie and Albert helped a lot and the church people. I kind of feel sorry for the church people. I really couldn't give my all to them. That sickness is nothing to mess with, shewee I never want that again! Well this is interesting what Auza showed me this;

According to a recent theory, the Lawrenceburg area is the likely site of "Chicasa"—the place where Spanish explorer Hernando de Soto and his men wintered in 1540-41 (earlier theories have suggested this campsite to have been in northern Mississippi. The Cherokee sold the area to the US in 1806.

Upon moving from East Tennessee in the early 19th century, around 1817, David Crockett served as a justice of the peace, a colonel of the militia, and a state representative. David Crockett established a powder mill on Shoal Creek originally called the Sycamore River. This area is now home to David Crockett State Park. Crockett was elected as a commissioner and served on the board that placed Lawrenceburg four miles west of the geographic center of Lawrence County. Crockett was opposed to the city located in its current location, largely out of fear of flooding. He and his family lived in Lawrenceburg for several years before moving to West Tennessee after a flood destroyed his mill.

After World War II, the Murray Ohio Manufacturing Company, U.S. producer of bicycles and outdoor equipment, moved its manufacturing operations to Lawrenceburg, building a new factory and assembly plant. Over the next several decades, the Murray factory grew to be one of the largest in the United States: 42.7 acres under roof.

I am so thankful that God is giving me some mountains and part of the Appalachins. I really got spoiled in Marion. This is a nice church that we are going to. These poor people have been without a pastor for almost two years! It is hard to find a pastor of a church. The church can't afford much and it is easier for a pastor to work and be pastor. Most churches don't want that. What about funerals, weddings,visiting the sick? Very hard to do. Auza had the grocery store but it was easy to find someone to take his place. We are moving into a brick shingled house.

From what we understand this church does have its troubles. There was a spat and members left and now it will be hard to get them to come back. Always something! We got to the house and Sister Stacy met us there and the deacon. Sister Stacy is a pleasant lady and very interesting. We got settled in and boy am I tired! The kids were hyper from sitting the whole trip. The church gave us a pounding. A pounding is everybody brings a pound of something, Really nice to have sugar and flour and beans. Auza loved the pound of peanuts! He had to run and get himself a pepsi with that. I made the kids some popcorn and they were happy with that.

As we got settled in, the kids went to "explore" the neighborhood. I guess they found quite a bit because they were gone long enough for me to put things away. I am so glad this place has a wash machine. We have a nice clothes line outside. Auza is at the church looking it over. Darlene and Zelma are sitting on the front porch. Someone gave them some chalk and they made a hopscotch on the sidewalk. Aneita and Glenda came back home and they played until dinner.

Albert and Merele found a nice park and played there for a while. When Auza came home he said it was a nice clean church with a nice size pulpit. The deacon said it will be hard to recruit for the church. He said with all of your kids we could get a couple more folks. I asked him if they have gone door to door lately? He said no not in a long time. I asked if there was enough in the treasury to buy some tracks? He said yes, so I went to the printer and showed them a track for us to give out. They will be printing them and I know I have at least six people to help me! He chuckled.

We went from door to door and asked everyone we could to church. That did bring in a few more people. Sister Stacy was very helpful in helping with the tracks. It is really nice here. After living in a farmhouse and Marion NC, it was mostly living alone away from neighbors. We are now a member of a neighborhood. Lawrenceburg had an Armory for the children to play. They had a big park with a big pool. They have fairs that came to town. There was a big statue of David Crocket. This is the place that my kids went to their first fair.

Their little eyes were full of entertainment. They saw clowns and rides and carnival acts. I wish I had a camera just to capture the looks on their faces. Aneita started acting like a clown and then Glenda followed. Robbie just giggled at everybody the whole time, Zelma was loving the carnival music going on the loudspeakers. Merele was everywhere, Albert wanted to watch the men hit the ball with a hammer that made the bell ding.There were clowns running around saying hi to the kids. There were rides and cotton candy. The kids had thought they had died and went to heaven!

I really loved the fair and we stayed here for 6 years. Aneita started first grade at 7. Glenda begged to go to school with her and I let her go. The teacher said "don't do that again" They are not like the school at the farmhouse. Glenda and Aneita played with a friend named Judy Beasley. She loved my girls and my husband! She was only 7 years old and loved Auza, we laughed at how she acted when she came to our house. She had a bike and that is how they learned to ride the bike. Once they learned they were on it all of the time. I felt sorry for Judy! But my girls were having fun and I let them.

On Sundays at the local park was a man that would come to the park and throw out a blanket. He would put piles of money all over the blanket. He would say when I blow the whistle you get as much money as you can get in your two hands. Their little hands grabbed onto that money and they thought that they were rich! Aneita came and gave her money to me and said, "buy yourself something pretty," with her eyes wide open. I said, "Wow, I will!" that made her happy. Oh by the way it was 50 cents in pennies and nickels. They all loved Sundays because of the money grab game. I wish they had that for adults!

Another thing that happened in Lawrenceburg was a place called the armory. The kids would go there on Saturdays and play all kinds of games and acrobats. When they came home from that place they were hungry, thirsty and tired! It did give me a break. This is probably the most exciting place to live. It is family centered and a growing city. Having six kids does not give you much time for yourself. I chose to clean the whole house every day. I send them off to school or outside and I clean every room. I don't want to have to fuss with them. It is a full time job, the only time "I" get is taking a bath, I lay back and let everything go! These are my thoughts as I relax.

If I had to decide which child is the best it would be all. If I had to choose which child was the smartest, I would choose Albert and Darlene. I think they had more calmness and time to observe more than the other kids had. If I had to choose which one was the funniest it would be between Zelma and Aneita. Zelma comes up with the funniest thoughts and Aneita acts out her funny thoughts! If I had to choose which child had the most tender heart I think that would be Glenda. She would come in crying if an ant got hurt. If I had to choose which one was the most creative that would be Merele. He can make something out of nothing and loves guitars! If I had to choose I would choose mine.

Chapter 15

Christmas in Cleveland

We are taking a week trip to Cleveland. We will visit Auza's mom and sisters. My mama and daddy have moved to Portsmouth Ohio to be close to all of the kids. Ola Mae still lives in Cleveland. I will go by the christian book store and see if Mr. Gardner is still there. We will go by the church and say hi. We will go by our old neighborhood. This will be our first Christmas with them . "Auza do we need to bring presents?" I asked. "Creolia, you know we can't afford that, how about you make some cookies?" suggested Auza. "Yeah, that would be good," I said with a sigh. "Now don't get yourself all stressed, they can't wait for us to get there! They will have everything perfect for everyone." he said with a smile.

Getting everything ready for six kids is a lot! Robbie and Zelma helped me. We have an old ford Nash. That picture that Auza saw in the newspaper had him on a mission. He willed and dealed with a man up the road and got a good price. His mama and sister helped him with it, thank God! I asked Auza how far was Cleveland and he said 10 hours away! 8 people in a car for 10 hours, Lord help us!

As we are getting into the car I said, "Aneita you and Glenda sit up front, Robbie and Albert get the doors. Do not let Merele sit next to Albert!" I am glad we have a bench seat up front. We weren't even ten miles from home and the fussing started. "Merele quit chomping your gum!" yelled Zelma. So he chomps it louder. I reach in back and make him spit it out. "Albert quit whistling," said Robbie. "Hey I know let's have a whistling contest," said Albert.

That wasn't such a bad idea we all had to whistle a tune and we all had to guess what song it was. There was another game we played, I spy. We had to do that quickly or it would be gone. Another car game put together a word. The word would be flower, you could get the letters off of a license plate or a billboard or a sign. We really loved that game! We would stop on the side of the street and eat lunch. I made plenty of bologna sandwiches and peanut butter sandwiches. I would have the kids switch seats so that everyone had a chance of sitting next to the doors. At night we would all fall asleep while Auza drove. Aneita would not go to sleep because she thought her daddy needed company.

As we pulled up to the house grandma Vinson's little head was watching for us the whole time. She is the sweetest lady I have ever met. They had a color tv and the kids laid on the floor to watch tv. Grandma Vinson saw that the girls legs were showing and she went and got the newspaper and threw it on their legs. The girls looked at her as if to say, "what are you doing?" I laughed. Grandma would sit there and whistle and lightly clapped her little hands. Eza and Mauga were making dinner in the kitchen, whispering. I went in there to see if I could help and they said no you just go on back in there and let us serve you.

Wow this is nice! They made us a wonderful baked chicken and mashed potatoes. We all ate like we were starving. Mauga kept asking, "Mer-e-le do you need anything? Auza do you need more coffee? Albert, do you want some butter with that bread?" She stood there the whole time and served us all!

Oh my, they had a christmas tree that was decorated beautifully! I could sit here and stare at that tree forever. They took a picture of the kids. Look how happy they were. The best kids in the world!

Mauga and Eza would go shopping at Woolworths and buy our christmas gifts. The girls favorite was a small bottle of perfume. The bottle looked like the top half of a girl with a lace blouse on.

They would get them all kinds of trinkets. They really outdid themselves! I guess they didn't want to get big stuff because there was no room in the car. They bought Auza some very nice guitar picks. They got me some Woolworths nylons, and some powder with the puff ball. My favorite was 5 hairnets! They may be small things but they are big to me. Albert and Merele got slingshots that kept them busy for a while outside. With Auza's new guitar picks they got out their instruments and sang christmas songs. The kids sat down and Eza made a copy of this song for them to sing.

Hark! the herald angels sing,

"Glory to the new-born King!

Peace on earth, and mercy mild,

God and sinners reconciled."

Joyful, all ye nations, rise,

Join the triumph of the skies;

With the angelic host proclaim,

"Christ is born in Bethlehem."

Hark! the herald angels sing,

"Glory to the new-born King!

Christ, by highest heaven adored:

Christ, the everlasting Lord;

Late in time behold him come,

Offspring of the favoured one.

Veiled in flesh, the Godhead see;

Hail, the incarnate Deity:

Pleased, as man, with men to dwell,

Jesus, our Emmanuel!

Hark! the herald angels sing,

"Glory to the new-born King!

Hail! the heaven-born

Prince of peace!

Hail! the Son of Righteousness!

Light and life to all he brings,

Risen with healing in his wings

Mild he lays his glory by,

Born that man no more may die:

Born to raise the son of earth,

Born to give them a second birth.

Hark! the herald angels sing,

"Glory to the new-born King !"

The kids sounded like angelic voices. I will hear them sing this song until the day I die. Christmas is so special and I am so glad we came to Cleveland.

As we were getting ready to say goodbye grandma was crying and the kids went and hugged her several times. Eza took Auza aside and gave him some money. They love their brother! As we were getting things together Mauga said, "come here children I have a surprise for you. We bought peppermint candy canes to have on your ride home." She handed them out. Then grandma wanted to say a blessing over us. It was the sweetest prayer, it went something like this;

"Dear Heavenly Father, we ask for your protecting angels to guide them on their way home. Let there be peace in their trip and no problems that shall arise. Be near them and hold them in the hollow of your hand. In Jesus' name we pray."We all hugged again and left with a swollen heart of love.

Chapter 16

A short stay

We stayed in Lawrenceburg for 6 years until a new pastor took over. There was a small church that needed a fill in for six months about 25 miles up the road. This was a short stay but a nice one. We lived up on a hill. There was a winding road in front. The kids would sit on the side of the hill and look at the license plates of the cars. They would see which cars came from far far away. There was this river that had trees alongside it. These trees had grape vines hanging from them. The kids would go there and swing on them. Albert and Merele were Tarzan. I could hear them from the house yelling that Tarzan yelled. The girls would look for pretty pebbles in the water and interesting rocks. They would collect pretty leaves in the woods. Sometimes they would stay all day out there.

The kids seem to have a lot of fun here. The time had come to move on. We were going back to where my mama and daddy lived. Sciotoville Ohio. We lived on a dirt road that had wet tar on them. It was like a holler but the houses were closer than most I have seen. We eventually moved to Mabert Rd. and that was a good place to live. We lived on a steep hill with about 20 steps to climb.

The winter was tough. Auza worked for Select Dairy to get extra income. The church we had, could not even support two people, much less 8! We had a very cold winter and we needed layers of clothes! The kids had to walk to school and it wasn't easy. Aneita and Glenda came back home several times because it was way too cold. Auza bought a bushel of apples for Christmas. We actually bought a tv! The kids were staying at the next door neighbors watching theirs, so Auza decided to buy us one.

I loved to watch Captain Kangaroo in the mornings. I watched with the kids until they left for school. 'Laugh in' was funny but a lot of skin exposed. The kids loved 'the Flintstones' at night. TV made a major change in our lifestyle. Instead of the kids running in and out the doors they sat still and watched tv. It actually brought a lot of laughter to our home. It also brought our home together as a family. It was actually very cozy.

There was this elderly couple that lived across the street. They loved my kids. They would give them treats and tell them spooky stories. It was really funny. One day Aneita had the hiccups. They started telling her a spooky story and she became so scared that she stopped hiccuping. They said Aneita where are your hiccups? They started laughing as she discovered they were gone. Glenda and Aneita loved going over to their house. Junior, my brother had a daughter named Debbie. She would come over to play. She was Aneita's age. Her mother had died.

One day Debbie was telling the elderly couple about not having a mother. They began to feel sorry for her and treated her extra special. Well this made Aneita jealous. Aneita got mad at her one day and said "the only reason people like you was because of your mama being dead." This made Debbie cry and when Auza found out that Aneita said that, well she got a spanking. She had to go and apologize to her. I don't think Debbie ever had a problem with her again.

We had a church and Auza bought some brand new song books for the church. It was just a couple of weeks and someone scribbled in all of the books. Our children were the only kids there a lot. Auza lined the kids up and said if the one who did this won't admit the scribbling then all of the children will be spanked. No one admitted it and they all got two spanks. It probably hurt Auza more than it hurt the kids. We never did find out who did it, but I do have my suspicions.

On Aneita's birthday I threw her her first birthday party. She is in the third grade and she seems to have a lot of friends. She had this one boy that liked her. I don't think she really knew it. She wasn't paying a lot of attention to him and he got mad at her. He pulled a really pretty ring out of his pocket and said, "I was going to give this to you so that we could go steady, now I have changed my mind!" He ran home. It did put a damper on the party because she missed out on a pretty ring. She also got a boulder from a guy "like I did" from playing marbles. There were a lot of "firsts" in this house but nothing prepared me for this one...

My mama and daddy came to church and to visit. When we all stood up to sing mama said "Creolia your daddy looked at me and smiled," and then he fell into the Isle with a heart attack! The church immediately started praying over him. We had to run to a neighbors house to find a phone to call the ambulance. I knelt beside my daddy and he was gone! I screamed and cried and all the whole church cried. "Oh God please please bring him back," I cried. My precious daddy had gone to heaven. Oh, I have never hurt so bad in my whole life.

We had the funeral in Katherines house, where they lived. His casket was put in there, and that is where he laid. That tall gentle giant lay there in peace. Daddy I will see you over yonder in the sweet by and by. I kissed his cold forehead with warm tears on his cheeks. I didn't want to leave his side. Katherine, Bertie, Lulubelle, Junior, Elvie, and Sherman were all by his side and mama's. It was not a good day at all, so so sad. You never know what tomorrow will bring.

I held onto mama when we went to drop his body in the grave, I kept hearing that song in my mind, "there ain't no grave gonna hold my body down. When I hear the trumpet sound, I'm going to get up out of that ground, for there ain't no grave going to hold my body down." Sleep tight daddy in peace for there will be one day we will meet again, over yonder.

Chapter 17

A Victorian House on a Hill

When I got home I had to spend time by myself and write down my thoughts; Daddy you no longer have to worry about them revenuers. You no longer have to work all day long. You now have the time to smell the flowers of the fields. Your days of labor are done. Daddy I will miss you and I will always hold you close to my heart. That big smile is engraved in my brain. Those big arms that would hug me, I still feel them. Yes, I will always be your baby girl. I will always have you in my heart. Tell everyone in heaven I said hi. Could you take your big arms and give Jesus a hug for me? I love you daddy, I love you. I closed my journal and sealed it with a kiss.

Mama and Lulabelle are in a daze and Katherine is tending to her. Mama said when she went to bed that night she felt Jared beside her. She said she even saw the bed have a dent from his body! I guess God knew she needed extra comfort. I have never had a loss like this and I don't want to! It was a dreary week, but time passes and things change.

Auza got another church. In Sciotoville. It was a small church but had a lot of possibilities. The church got us a beautiful victorian home on a hill with three pear trees and an apple tree. It had a big huge front porch. We are preparing to move and then we got another bad news. My brother Sherman was working in his yard and his tractor rolled backwards and covered him up under the mud. Oh my Lord, his poor family and small children. What in the world are they going to do?

It was too sad to mention! Oh my heart ached for them! I did all I could do. I cooked and tried to help. What can you say or do, nothing will bring him back. It seemed like yesterday he was getting ready for his first date. With his dark hair and looking so good. His big smile looked just like daddy's. His big heart and love for his mama and daddy. "Sherman I love you, I will miss you. Be at peace my brother and I hope to see you in the sweet by and by. I kissed his cold forehead with warm tears flowing down on his shirt.

We returned home in silence and packed everything and moved. This house is the prettiest house we have ever had. I could really like this after all the sadness in the last couple of months. I set everything up so pretty. The kids are out meeting new friends. They are all eating from pear trees and apple trees. The front porch is so big I can hear them running from here to there.

"Mama, do you have some clothes that we can play dress up?" asked Zelma. "I will go and see," I said. I found some high heels and fancy dresses that I will never wear. I took them out to them. Oh my it was so funny! They were acting like they were movie stars. Zelma sanged that whole song "sad movies always make me cry." If I had a movie camera she would have taken the show!

Clonk, clonk, clonk, all day long their heels on the front porch. I don't know what the boys did but the girls had a show! Even the neighbor girls came over and joined. Auza brought home a bushel of fresh green beans.

We went and sat out under a big oak tree and stringed the beans. I guess I make the best stringed beans, because even the kids had seconds. I make some pinto beans the next day and cornbread, with green onions and tomatoes. Again it was a big hit! I guess cooking is therapy for me. I love to cook and above that I love to eat my own cooking, I chuckled to myself.

They have the Grand Ole Opry on tv and I love to hear Loretta Lyn and Dolly Parton, Flat and Scruggs. Their singing reminds me of the hills. Gospel is my favorite above all. Me and Auza sing in church together. We go there and practice once a week. Zelma and Darlene have been singing. I am glad they like gospel music. When we get to church early all the kids of the church play Red Rover. These kids have no worries except their next meal. I understand the bible saying become like a child. Children don't worry or fret . They are not anxious for tomorrow or what they will wear or eat. We need to become that way like a child. This world would have a lot less worries!

I still like Auza's preaching, he always has something new to say. He said this past Sunday that Daniel was in the lion's den and now he is playing with the lions. David had slain a bear and now he plays with the bears. Moses saw it rain manna from the heavens and now he is with the manna, Jesus Christ! Boy what a thought! Portsmouth Ohio was good and kind of a struggle, but more fun than struggles. The kids met a lot of friends and really hated to go. Poor things have gotten used to all of this moving. I guess it's like a military home. They move here and there all the time. Memories no matter where you go. Just make them good memories!

My sister Bertie had two sets of twins. Her children were much older than mine. This picture was taken in 1966. This was my children's ages were,when we moved to Toledo in 1966;

Albert 17   Zelma 13

Robbie 15   Aneita 11

Merele 14   Glenda 10

Bertie was a vibrant lady and great to be around. She spoke her mind and was funny to the point of her honesty. Bertie loved to cross stitch and quilting. Later on in years after the kids were grown Bertie and I made a quilt of dolls of the world. Bertie quilted and I made the dolls. This quilt was huge! It was at least a king size. All the dolls of the world I put on them. Aneita bought the blanket for $200. We were proud of our work that we did. I hope Aneita will enjoy the works of the Gifford hands!

Bertie and Katherine live in Portsmouth. Katherine's kids were 2 boys and 2 girls. Her two sons were the age of Albert and Merele. Her two daughters were the age of Glenda and Aneita. Aneita spent a week with her cousin Gail. Up in the hollers. They went to summer school and the school was like a small church. Aneita loved her cousin Gail. We always will stay in contact with my family as we go forward to another adventure.

Katherine and her husband worked in the church together. Donzo was a wonderful preacher. They would sing and preach, it was a gospel hour with them! Junior came to our church in Sciotoville. Elvie we never saw him much. Lulabelle ended up marrying a man in Portsmouth and they were so cute! Lullabelle had the prettiest kids! She was a good mom and wife.

This is us in 1966 from left to right back row Robbie,Merele,Zelma,Albert,Crolia,Aneita,Glenda,Auza

This is Bertie and Chester and their children in 1966. From left to right, Licille, Janet, Earnest, Edith, and Edna

This is the quilt that Bertie and I made.

We went into Highland Bend and said goodbye to mama. She was out milking the cow. She had her bucket full of milk and took her finger and swiped the cream off of the top. I wish I had a camera at that moment. She then sat down in her rocking chair and asked Aneita to comb her long grey hair. She didn't really act sad, happy or not really any emotion. I guess she has plenty of time to reminisce and is at peace with herself. We had the car packed and the trucks ready to go to Toledo Ohio! Goodbye mama come and see us sometimes. She smiled and waved goodbye. This was a picture of us during that time. Look at my smile, yes I was very happy. I love my family and will always cherish the moments together.

Chapter 18

The Buckeye State and City of the Mud Hens

Toledo Ohio and here we come. The kids were fascinated with the tall building. It was 1967 and I know nothing about Toledo ohio. Auza got this photo at the library. This is the downtown area of Toledo Ohio.

Early Toledo

This is the biggest church we have ever had. I guess the membership is 100 people. We are moving on the westside of Toledo. The house is a two story with 3 bedrooms and one bathroom. The one bathroom is big, thank the Lord! A really nice neighborhood with a nice fence in the backyard. Clean I would say. I hope these neighbors don't have a fit seeing a family of eight living here. The kids were all excited and the neighbors welcomed us here. There was a policeman that lived across the street. A widow woman next door raising her sons. I don't know but I think this will work!

My children's ages were when we moved to Toledo

Albert 17   Zelma 13

Robbie 15   Aneita 11

Merele 14   Glenda 10

Albert had signed up for the service. He is going to be a Marine. They will be sending him to San Diego Ca. Camp Penalton. Albert had a best friend in Sciotoville whose name was Jerry. Junior had married this lady with children. Jerry was the same age as Albert. They would do everything together. So when Jerry signed up for the Marines, that made Albert all the more motivated.
Unfortunately he was sent to Da Nang, the roughest area of all. It was 1968 Christmas eve when Albert was to come home. President Johnson put a freeze on anyone coming home. Albert got shot on Christmas Eve. A man that was in his troop and from Toledo Ohio carried him to the base.

When I saw three dressed soldiers walk up to my house, my legs gave out on me. It was the worst news ever heard in my life. Nothing, nothing can compare in losing your child! Aneita had just sent him some peppermint candy canes for Christmas. Right before he died he begged Auza to get him out. Auza had called the government officials and because he was not the only son, they would not send him back. They never sent him back until 3 weeks later. They sent him in this dented tin can casket, it was awful. Auza called the government and requested a nice casket and they did. Oh my Lord My God my heart aches and I hurt all over. I will never be the same. I went around for months not wanting to talk to anyone, I needed some time.

Toledo was a big city almost like Cleveland. People here sweep their sidewalks and nice yards. It was really nice. Robbie and Zelma joined a gospel group at church. They were really good, I was so proud of them! Aneita and Glenda had bonded with kids their age. Merele was always up to something. I guess you could say he was pretty independent. In 1969 we had the best Christmas ever. Look what Auza bought the whole family. He got us a fancy tv and a big beautiful picture for the wall. This was really nice!

One of the girls bought some jax. I surprised them with my talent. I knew how to flip the ball up and swipe a jax. They have around the world jax, pigs in the blanket jax, kiss the jax, oh my I can't do all of that. We are happy most of the time. I think the dinner table keeps us all together. We had a big living room. Two chairs and a long couch. We would all laugh at Carol Burnet, my Lord where does she come up with all of her comedy. I laugh until I cry. Auza loved the Red Skelton Show. He was a natural that's for sure! The kids loved laugh-in. It gives you something to look forward to! Dark Shadows with Barnabus! Peyton Place with Ryan O'Neal.

This tv lasted for years!

Ok here are the programs we watched in 1970. Bewitched, My three sons, The Munsters, The Adams family, Truth or consequences,The newlywed game,Dark Shadows, General hospital, Bozo the clown,Bananza,Green acres, Captain Kangaroo,Candid camera (Auza's favorite),Mike Douglas,Johnny Carson,Ed Sulivan show,Carol Burnet (My favorite) just to name a few.To adjust the color on this tv we had to walk up to the tv and work with the knobs and antenna. It turns green and everybody looks like a walking pickle. We never can get it perfect. I must say it was perfect for watching the Munsters.

The kids had this park that they went to called Joey Brown Park. Aneita loved it there. There was a pool and a ballpark. I guess it was a hangout,she met this girl from Poland, who couldn't barely speak english. But they hung out with each other. Glenda had a friend up on the busy street and her and Aneita finally split and went their own ways. It seemed everybody found themselves. We were all comfortable. Everybody lived in this house until they all went their own ways. I eventually got a job at a vacuum factory. That gave us enough money to take a train trip to Colorado. We found places for the kids to stay. This was the adventure of my life!

I love the sound of the trains and the smell of the smokestack. The views are just unbelievable! When you are riding on a train the world looks much different! The tracks are all in the forest where nature abounds. The bed is not that comfortable but the movement of the train rocks you to sleep. The food and coffee is really good! They come up to you in the train suit with a white cloth napkin hanging off of their arm. Very nice!

I have seen the mountains in the Appalatians but those were hills! Colorado has mountains! Their mountains are so high that no trees grow to a certain point. There are snowpeak mountains in July!

Tour Guide said,

Snow peak mountains, Lush green forests, natural waterways and reservoirs, Grand canyons and deserts, Colorado offers many delightful experiences that can leave you mesmerized. Imagine if you were to enjoy these natural beauties in the middle of a forest. Imagine you were a part of a scenic train ride in Colorado immersing yourself in all the fascinating stories and glimpses this historic state has to offer? Take these various routes and train journeys to identify and visit the Hogwarts of your dreams. Listed below are some of the scenic train rides in Colorado. This historic coal-fired steam engine will gently chug you back to the industrial revolution days through the lush green meadows and steep mountain canyons. Offering you a rare western scenery, this ride, together with its coaches with a Victorian elegance of the parlor car or coach car, will take you on this journey to Colorado. This historic journey also includes a delicious buffet meal.

Seal that with a kiss because that is exactly and more beautiful than you can imagine. I hope Heaven has rolling hills with snow peaks. I will take my wings of flight to the top and praise my God! When you see all that He has created it kind of gets you all excited about the sweet by and by. A treasure to always remember and a place to go back to. An adventure truly made for the soul! I think one of my favorite things to do was getting off the train and walking the cute little towns. I think I could do this for a living!

The conductor would come to our tables and tell us about the next stop. The food they served was like a Bird of Paradise sitting in the middle of our plates. Every dish had a special design. The bread was made there and the butter. They had homemade jam. He asked me, "how do you want your steak done?" and I said, "cooked." Auza laughed and said, "Do people eat their steaks any other way?" I still don't know what they expected me to say. People would say hi and very nice. The only thing I wanted after eating was to look at the view from my window.

There was one view I saw and it was a mama bear with its baby cubs. I said, "Auza, Auza look at this!" He smiled and said, "mama bear said the portage is too hot." Another site was Deer and antelopes running together. They look like they have the best life. When we went to one of the small towns right before Colorado it had a western show. When we first got there two men were in the middle of this dirt road. They were drawing their guns. Suddenly the sheriff came in on a horse and they shot at him but he got them both. The whole town was dressed up in western apparel. I bought a folding fan and Auza got a sheriff's badge.

When we finally got to Colorado it was Colorado Springs and oh what a tour there was. Clear crystal water streams. Caves with water and ice cycle looking things. We took this rugged old bus that took us all the way to the top of the mountain. Oh my I did not think it was going to make it! When we got to the top it was breathtaking, but it did make my feet hurt! When I am up high in a place my feet hurt. Maybe it has to do with those hot bricks the women put on my feet when I was sick.

Chapter 19

When I hear that whistle blowing

I wrote how I felt about that trip on the train but I feel like I need to share how much this trip meant to Auza. Auza was from Crumb WV before they moved to Cleveland. They only had flour and water at times to eat. Auza was fascinated with the trains there in WV. I guess in his imagination as a young child maybe 12-13 years old he decided to start hopping trains and see if he could find something outside of Crumb West Virginia for his mama and sisters. When he started hopping trains he met all kinds of people and he accomplished a lot to get some food for his family that's why trains mean a lot to him they saved their lives.

I can't imagine how he felt with two sisters and a mother hungry and he had all that energy and time to maybe be a help and I'm sure that's what his trips consist of and also some adventure with that. The stories he could tell you of trains and how symbolic they are in his life. Riding in a train going all the way to Colorado with that steam engine, his eyes, his ears, his heart and his soul was in that train. I would not doubt it a bit that God the Father will have a train to pick him up and take him to Heaven when he dies, that's how much he loved the train.

I know every single time I hear that whistle blow now I will hear "All aboard!" Happiness is not in what you have it is in what you do. A grateful heart is always content in the present situation. God allowed us to take time and smell the smokestack. Our life has been through a lot and this was a reward from God. Praise him in good times and bad. Never underestimate The Father!

When we got home it was back to work and home life again. The trip did put a spring in our step. The kids were asking about the trip and of course I went on and on about the beauty to behold. It seemed that the kids had a good time but were ready to come home. "Ok what for dinner tonight? It will be up to you," I said. "Hamburgers and french fries" they said. "Oh, that will be easy to do." I said with a smile. I make my burgers with soy sauce and onion seasoning. A nice slice of onion on top with a fresh ripe tomato and lettuce. There is nothing like a good ole burger. Shewee! That filled me up, time for a nap!

It is really hard to come from such a great trip and beauty and relaxation. I don't want to work, I don't want to do anything, my mind is still on vacation. I think I will take a nap and maybe I can snap out of it. I closed my eyes and I fell asleep hearing that whistle blow and the steam coming out of the top of the train. "Daddy what are you doing? You have your suitcase in your hand. Where are you going daddy?" "Creolia I am waiting on that train to take me home." "But daddy you are home, right here is your home!" "No, my home is over yonder." "No daddy, your home is right here, please don't go, stay here!" I was tugging him toward me. "Baby girl I have to go and it will all be alright." He stepped into that train. "Daddy I love you!" I cried. I woke up crying and missing my daddy so bad.

Strange things happen and you don't know why and the purpose of it all. I am sure mama misses him terribly. Oh my I need to get my head together and start getting busy with something. Look at all of this laundry all over the floor, those kids just throw things anywhere, I put the clothes in the washer as I complained. I hear the girls practicing a song for Sunday. Zelma has a beautiful voice. I started humming the song that they are singing. "Shackled by a heavy burden, beneath the load of guilt and shame, then the hand of Jesus touched me and now I am no longer the same. He touched me, and oh the joy within my soul. Something happened and now I know, he touched me and made me whole." Hallelujah! I feel like dancing, I chuckled to myself! Well that took my blues away!

As we got up it was a Saturday and everybody was home. The house is ringing with the noise of everyone getting ready for the day. The first thing I do is get Auza his hot coffee while he gets the newspaper. I then put bacon in my iron skillet and biscuits in the oven. I'm cooking and Auza came in and said that there is a tornado watch in our area. It did look gloomy outside.

Auza turns the news on and it keeps getting worse. They now have spotted a tornado. We really did not know what to do but watch. We walked out on the front porch and suddenly everything got eerily quiet. No birds, no ruffling of the trees, not a sound!

Suddenly a distant noise and we all ran inside and looked out the window. Like a rushing wind it came and the street became a river. Poles fell down, trees were knocked down. It was the scariest thing I have ever seen. The news was warning that there was live wire in the streets. Please do not walk into the water, it is full of electricity. Well Zelma had a close friend that took that dare and got electricuted.It broke our hearts especially hers, That was a tragic and sad day in Toledo Ohio.

See how quickly things can change. We woke up for the day. All spry and happy. Ready to eat and enjoy the weekend. Then Boom! Things changed just like that. You can plan your day but don't plan on it.For I know who holds tomorrow and I know who holds my hand, so the song goes. The day was quiet, it seemed we all were in shock. It made me cherish my family even more. We had popcorn that night and watched a movie. We really didn't want to do anything.

There wasn't any church on Sunday. Wires were still on the streets and trees down. Things were quiet in the neighborhood. Today would be a good day to throw a roast in the oven. As I prepared the roast I could hear the kids laughing really loud. I walked into the living room and they were watching, I love Lucy. I stood there and laughed with them, Lucy is so funny, she reminds me of my sister Bertie. Ethel is just as funny! This is going to be a good day.

It felt strange to have Sunday to ourselves.I don't think there has been a time since I've been married to Auza other than being sick that I missed church on Sunday. This was an exception with all the tornadoes and storms,it's probably going to take a week or two to get the streets cleared up I suppose.This is a great time to just relax and enjoy my family and enjoy eating at the dinner table and having some laughter and fun being home.

I went and took a nice long bath. I got my Jergans lotion and smoothed my face and legs with it. I put my nightgown on and braided my hair. I spent some extra time on myself. I have been getting some magazines in the mail and I think I will read my Southern Living. It is a good day to be alive. You know sometimes life just flows!

Chapter 20

Creolia's swinging kitchen door

One thing that we all have in common and that is food.Aneita's friends, Zelma,Darlene,Glenda's and sometimes Merele's friends came to eat. Especially if they spent the night they "expected" my gravy and biscuits. My gravy and biscuits weren't just gravy and biscuits I had fried potatoes I had bacon, I had sliced tomatoes, homemade jam on the table, homemade butter the whole works that's what they got! The smell alone would drive you nuts and afterwards we all would go somewhere and flop down and watch tv or whatever.

If unexpected people come at lunchtime to my kitchen most likely see me making, fried potatoes, sliced tomatoes, sweet onion, fried sliced bologna, bread, mayonnaise, mustard. Sometimes if I had biscuits left over from breakfast I would make sausage biscuits for lunch. Lunch to me was a mood thing.If it is sunny outside and the windows open, I might make american hotdogs or american hamburgers. If it is rainy homemade soup! But if nothing was happening fried bologna.

The dinner bell is ringing! Stuffed green pepper, meatloaf, fried chicken, cabbage and sausage, cooked greens, a good rump roast with carrots and potatoes and gravy, Stipped steak and peppers and noodles, pot of fresh green beans, chili,bean soup (pinto, butter beans, white northern with a ham hock, black eyed peas). Spaghetti (I did get it mastered!) and salad with homemade dressing! Cole slaw with BBQ ribs, I guess I can go on and on.

No complaints from the family about my cooking. That was how I gave them my love everyday! I would sometimes be home by myself and I would get on my knees in my bedroom and cry out to God for my family. Tears flowing from my eyes asking for his blessings. There were times someone came home and I am sure they heard me...which is good. I have so much to be grateful for and praise his name everyday. I even prayed for my children's friends! Someday I will go to heaven and we will have a feast with the Father and his son. That will be God's swinging kitchen door!

I can only imagine what that table would look like. There would be a long long table as far as you could see. People from all around the world feasting on the food The Father has to give. I can imagine that the angelic host will serve us. I can only imagine big beautiful flowered bowls on the tables filled with fruits that we have never seen before. Most likely the food that we will eat we have never tasted before. As we drink the fruit of the vine in our pure golden goblets. The angelic choir singing songs of glory to God on high. I can only imagine Jesus sitting at the end of the table and presenting a toast to all those that had made it. We will be sitting in the clouds eating, singing, praising and worshipping God our Father and his son Jesus Christ.

You know I am ready if he should call. But while I am here on this earth I will do my best. As a mother I gave my all in raising my children. I had no experience. I winged it many times. My main goal was to have a hot meal on the table. Clean clothes for my family to wear. A clean house to live in. With all of that, that did take all of my time. Having six children and all that is required of you, that doesn't give you much time "have fun." I find myself reminiscing of the times I had before the kids. Okay I am going to reveal a secret with Ola Mae and me.

There was one day at school that Ola Mae was really having a bad day. I wasn't feeling well either, my daddy was in the hospital and a new job. I didn't have my homework ready. We had some spare change in our pockets. So we took the bus to the zoo. We ran around like doves out of a cage. We made faces to the monkeys, danced in front of the gorillas, and growled at the lions. We were in wonderland! We bought a box of cracker jacks, we traded our prizes ha! They had a bubble gum machine and I got a round purple one and she got an orange one. Chomping on our gum and then we sang a jump rope song.

"My mama told me if I be goody that she would buy me a rubber dolly. My uncle told her I kissed a soldier now she won't buy me a ruby dolly.

Chewing chewing gum, chewing chewing gum,chewing chewing gum, and there we stayed all day! We laughed as we sang that song and chomped our gum really loud. When I think of that it always makes me laugh.

Another time we misbehaved it was involving my friend Freddy. Freddy was getting me a pepsi to drink at the roller rink. Olla Mae said "Come on Creolia and let's hide from him." we had our skates on and clumsy trying to walk on the carpet. We went behind a bar where they sold food, the person wasn't there. We were kneeling down and we heard a voice that said "Maam, umm ma'am. I peaked up and it was Freddy, who was going to order food. Ola Mae and I popped up and said "how can we help you sir?" His eyes opened so wide and said "what are you doing?" We started laughing and ran back on the skating rink. He didn't find any humor in that. Always reminiscing and thinking of the fun times in my life.

The one joy that I do bring to the table is my food. Everybody likes Creolia's cooking. So being that I am proud of what I make, I made for you some recipes to enjoy! I will see you at that big dinner table that God has prepared for us! Praise be to God I will get to be served my food!

Cheers until we meet again!

Creolia's Hotdogs

1 can of Hormel Chili

\1 Pack of Ballground hotdogs

Sweet relish

Chopped Vadalia, Shredded Coby Cheese, Cole slaw, Spiced mustard and ketchup

Preparation

To prepare a good hot dog doesn't take much talent, it's all according to what you prefer in your taste. Take a pack of ballpark or whatever brand name you want hot dog and I like to put mine on the grill. Have the grill already hot with charcoal so your hot dog will have that charcoal flavor. When you put it on the hot grill, close the lid immediately because if you don't the hot dog will not cook inside only the outside. I like to make my hot dog dark and I like the skin to be a little crispy. You can take the hot dog buns and open them up and toast them right before you put your hot dog on your bun. Put whatever dressing you would like but I like the chili, sweet relish and Vidalia onions,cole slaw on mine with shredded cheese.

Creolia's Potato Salad

Ingredients

5lb of Yukon gold potatoes

1 Med Vidalia onion diced

One stalk of celery diced

2 Kosher dills diced

5 boiled eggs 3 diced and 2 sliced for toppers

3 tbsp sweet relish

1 1/2 Cups Miracle whip

(Miracle whip has more flavor than mayo) ¼ cup of Dijon mustard
1 1/2 tbsp of sea salt (sea salt has a more salty taste), 1 tbsp pepper, 2 tbsp paprika

To prepare your potatoes, put them in ice cold water. Do not peel them, the peeling seals the firmness of your potato. When it comes to a boil set your pot aside and let it get to room temperature, take the potatoes out and let them dry. Prepare your ingredients for the potato salad as the recipe shows. Delish.

Creolia's Macaroni & Cheese Taco Salad

Ingredients

1 lb of hamburger

1 pack of McCormick taco seasoning.

1 box of hamburger helper mac and cheese

1 small Vidalia onion diced

1 med. tomato diced

½ cup of original rotel

1 cup of green and black olives sliced, 1 cup of shredded sharp cheddar cheese, Chives diced, 1 cup of sour cream, 1 bag of tortilla chips.

Preparation

You will take the hamburger and put it in a frying pan. After you have cooked the hamburger drain and the add taco seasoning sauce packet, McCormick is the best in my opinion. Add the Hamburger Helper mix in with it. Once that cooks you will mix in your diced tomatoes, Rotel, your diced onions into the mixture and half of your olives into the mixture, mix it well. Get a large platter and pour it in the middle of the platter. Put olives and sour cream and cheese on the top. You will take your tortilla chips and trim underneath the setting of the dish and serve it like a finger food.

Creolia's Chicken & Dumplings

Ingredients

One whole small chicken

5 cups of water

3 large carrots diced

1 large Valdalia onions diced

Frozen peas

3 cups of self rising flour

½ cup of milk

1 stick of real butter softened

¼ cup of parsley flakes, 2 Tbsp Italseason, 2Tbsp pepper, 2 Tbsp sea salt

Preparation

First thing you will do on this recipe is get that whole chicken with skin and all into a crockpot and cook it all day. That will bring in your broth and will give you bone marrow, this truly is the best chicken flavor you can get. After you have cooked your chicken, remove the bones and the skin and put all the meat back into the broth. You then add your diced carrots and diced onions, while they cook take your flour and milk and knead it with your hands. Add the softened butter and the parsley flakes, salt and pepper and Italian seasoning in your dumplings. Once you have kneaded the dumplings you will take it and roll it out with a pin roller and make thin square dumplings, remember this is self-rising flour. Take the dumplings and drop them into the mixture and they will cook for about 20 minutes and when they start to sink they are done. Add your frozen peas the last 10 minutes.

Creolia's Green Beans

Ingredients

6 thick sliced bacon

3 lbs of stringed green beans

A large vidalia onion

Two large potatoes or four

Salt and pepper

½ stick of butter

Preparation

Dice your bacon and string your green beans. Put your beans in ice cold water and get your water come to a boil. Put your diced bacon and onion into the water. Make sure the water covers the beans. Uncovered cook on medium for several hours. After two hours put your potatoes in, let them cook for another half hour. Taste your bean. If it needs more salt then add because the bacon adds salt. Put some pepper in. Add your half of a stick of butter at the end to add sweetness. Should be good to go!

Creolia's Fried Chicken

Ingredients

4lb of chicken

3 cups of self rising flour

3 eggs

½ cup of half and half or buttermilk

2 Tbsp paprika
2 Tbsp salt
2 Tbsp pepper
Butter flavor Crisco Oil

Preparation

The first thing that you do in making good fried chicken is to thaw your chicken out well and pat it dry, right before coating. Get a bowl and put your egg mixture in the bowl with the milk, paprika, salt and pepper and mix it well. Get a large paper sandwich bag or plastic bag and put your 3 cups of self rising flour in it. Take your chicken, dip it in the bowl of the liquid, put it in the paper bag, shake it then take it and dip it in the bowl again put it in the paper bag and shake It and put it in the oil at 350 degrees .

Creolia's American Burger

Ingredients

2 lb of hamburger

¼ cup Soy sauce

2 Tbsp onion season

2 Tbsp pepper

Cheddar cheese

Tomatoes, Thinly sliced dill pickles, Sliced Vidalia onions

Lettuce, Mustard, Ketchup,

Preparation

You take your 2 lb of hamburger and knead it with your hands until it's totally soft. You then add your soy sauce and your onion seasoning and your pepper. Mix it into the hamburg then pat them out firmly and into your skillet. Have some oil in the skillet on about 350 and that way immediately sears your juice into your burger. You do not turn your burger over until the blood is running out the top of it. You then turn it over gently, do not press it! Pat gently that way it will keep a juicy burger. Let It sear on the other side and 10 minutes it is done. Toast the inside of your buns.

Creolia's Biscuits

Ingredients

2 1/2 cups of self rising flour

1 cup of milk or butter milk

½ stick of butter cold sliced

½ stick of butter whole

¼ tsp of salt

Preparation

If you have an iron skillet, put a half stick of butter and put it into the oven on 350. Put your flour into a bowl and the sliced butter in and slowly knead the flour with your hands. Slowly pour milk in to your liking. Put a dash of salt in. Roll your biscuit mix out and get a drinking glass and make your biscuit cut out. Get the iron skillet out of the oven and take your biscuit and turn it over on both sides in the skillet, so that both sides get butter. Put into the oven to bake until golden brown.

Creolia's Gravy

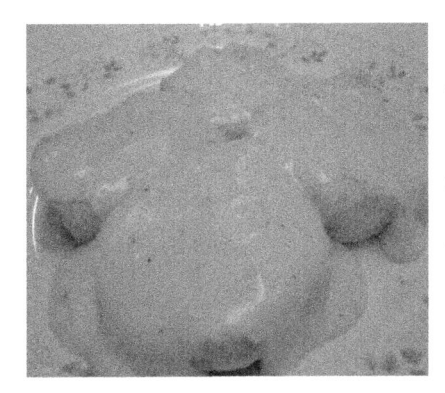

Ingredients

½ lb of thick sliced bacon

⅓ cup of flour

3 -4 cups of milk

Pepper

Preparation

In an iron skillet fry up ½ lb of bacon. Take the bacon out and while the grease is hot put into it ⅓ cup of flour. <u>You will cook the grease and flour until it turns a dark golden brown.</u> Slowly stir in your milk while stirring constantly. Stir your gray until it becomes thick. Put a dash of pepper.

Creolia's Coleslaw

Ingredients

1 head of cabbage

1 large carrot

½ cup of Miracle Whip

¼ cup of vinegar

½ cup of milk

¼ cup of sugar

Preparation

Shred a head of cabbage In a a bowl mix the ingredients well
Pour it into the cabbage and stir Refrigerate for a hour

Creolia's Roast

Ingredients

1 cup water beef sirloin roast

½ cup of flour

1 pkg of beef lipton onion soup

1 tsp garlic powder

2 Tbsp of parsley

1 Tbsp pepper and salt

1 Large onion sliced thick

7 medium sized potatoes,1 pkg carrots

Preparation

Preheat the oven to 350F. Sprinkle 1 cup of flour and beef lipton on the bottom of a large roasting pan. Add water and stir. Place the roast in a large roasting pan on top of the flour mixture and sprinkle roast on all sides with spices. Rinse potatoes and onion, chop into quarters. Arrange potatoes and onion around roast along with carrots Bake in a preheated oven according to the weight. At least an hour and a half, if you put it in while at church put it at 250.

Tips-Every once in a while take the juices and coat the roast and vegetables

Creolia's Pinto Beans Soup

A bag of pinto beans

2 Ham hocks or 5 slices of bacon

½ stick of butter

Chopped vidalia onion (for topping)

Salt and pepper

Cornbread (2 cups cornmeal, ½ stick of butter, ½ cup of milk , 2 eggs, 3 tbsp of sugar, dash of salt

Take your pinto beans and you put them in a strainer and rinse them well to make sure there's no rocks. After that, put it in a pot and cover them with cold water with a little salt overnight. The next morning put them on the stove and bring them to a boil. Then what you're going to do is put the ham hock in your beans at the bottom of the pot so the ham hock cooks well and for the next two to three hours you will be simmering your beans until they're done. If the water evaporates just keep adding water to cover the beans. I like a soupy bean, not a dry bean. When your pinto beans are finished put in a half a stick of butter that gives it a really pretty shiny look and also sweetener.

TipsThe cornbread, mix the ingredients. Have a cast iron skillet in the oven coated with butter on 350. When the skillet gets to 350 pour your mix in the skillet (this gives it a crispy bottom) Bake until golden brown about a half

Creolia's Fried Okra

Ingredients

3/4 c. cornmeal

1 ½ cups of

1 tsp. paprika

1 tsp. Black pepper

1 1/4 c. buttermilk

A large bag of okra

Preparation

Add all dry ingredients in a bowl and mix. Put buttermilk in a bowl. Put your okra in the buttermilk. Then put it into the dry ingredients. Have your skillet hot on medium heat. About an inch of vegetable frying oil. Cook for 8-10 minutes.

Creolia's cooked greens.

1 smoked ham hock or 4 slices of bacon fried

2 tbsp. sugar

1 med. vidalia onion diced

½ cup of turnip or parsnip

½ cup sorghum

2 cups of collards and turnip greens

1 tsp of pepper and salt

Preparation

Sear your ham hock in the iron skillet or fried bacon. While that cooks put your chopped greens in a large pot. Cover with water and bring to a boil. Add your ham hock and juices into the greens. If frying bacon put the grease in and when the greens are done put crispy bacon on top. Add ingredients into the mix and simmer for an hour and a half. Add the butter the last 15 minutes.Simmer with the lid off so that the water evaporates.Tip;You can transfer ingredients into a large iron skillet.

Creolia's Blackeyed Peas

Ingredients

1 pound (453grams) black eyed peas
4 -5 thick bacon slices , chopped
1 cup smoked sausage or turkey , diced
1 large onion , diced
2-3 teaspoons minced garlic
2 teaspoons fresh thyme, finely chopped
bay leaf
1-2 teaspoons creole seasoning
7-8 cups of water,2 cups or more Collar greens , 1`/2 stick of butter,Salt and pepper to taste

Preparation

Rinse dry black-eyed pea beans and pick through and discard any foreign object. (I did not have to do this because I used the package beans,). Add beans to a large pot covered with 3-4 inches of cold water. Cover and let sit for about 2-3 hours.In a large, heavy pan, fry chopped bacon until brown and crispy for about 4-5 minutes . Remove bacon mixture, set aside.Throw in the onions,garlic, thyme and bay leaf and saute for about 3-5 minutes, until onions are wilted and aromatic. Drain the soaked beans, rinse, and place the beans in the pot. Season with creole seasoning and salt to taste. Mix and bring to a boil.Reduce heat to a simmer and cook, uncovered, for about 20 minutes.Throw in the collard greens, and bacon into the pot, continue cooking for another 10 minutes or more, stirring occasionally, or until beans are tender and slightly thickened to your desire.Add more water if the mixture becomes dry and thick, the texture of the beans should be thick, somewhat creamy but not watery.Remove the bay leaves.Taste and adjust for seasonings with pepper, creole seasoning and salt and butter as needed. Serve over cooked rice and garnish with green onion.

Creolia's Chili

Ingredients

2 Lbs of hamburger
1lb of hot sausage
1 medium green bell pepper, diced
1 large onion, diced
2 large cans peeled whole tomatoes, blended
2 cans of Bush baked beans
2 cans dark red kidney beans, drained
1 can of tomato past
1-2 crushed and diced cloves of garlic or 1 Tbsp of garlic powder,¼ cuof chili powder, 2 Tbsp paprika,1 teaspoon crushed red pepper,to taste, salt and pepper

Preparation

Combine the hot sausage and the beef with salt and pepper, and brown over medium-high heat in a 5-6 quart dutch oven or stew pot. Add the onion and green pepper, and cook for another 5 minutes, or until the vegetables are slightly softened. Add the remaining ingredients and bring to a boil. Reduce heat to low-med. and simmer for 60-90 minutes, stirring about every 15 minutes. Serve immediately or refrigerate for 24 hours for a more complete flavor. Top with shredded cheddar, crackers, or your favorite chili topping. Tip: Add a ½ stick of butter when it is finished, that adds a shine and sweetness to the chili.

Creolia's Stuffed Pepper

½ cup uncooked white rice
6 eaches green bell peppers, tops and seeds removed
1 pound ground beef chuck
½ pound ground pork sausage
1 med vidalia onion
1 egg
salt and ground black pepper
1 can of condensed tomato for topping

Preparation

Place rice in a small bowl and add warm water to cover. Soak for 15 minutes. Drain and transfer rice to a large bowl.**Step 2** Add ground beef, ground pork, onion, egg salt, and pepper to the rice. Mix by hand until all ingredients are well combined. Stuff beef mixture into bell peppers and place in a Dutch oven, with stuffing facing up. Pour tomato juice over the top. Bring to a boil over medium-high heat, about 5 minutes. Reduce heat to medium-low, cover, and cook until peppers are tender and meat TIP: filling is cooked through, about 1 1/2 hours. Baking in the oven is good also.

ME!

Our trip to Colorado

Alberts school on the farm. The boy in the striped shirt, in front on the left.

Specially-Built Delivery Truck
Caters to Lumber Retailers

Auza worked for a lumber company in Lawrenceburg Tn. He made the news!

Me and Auza are on the left side, I have a white coat on and Auza a suit.

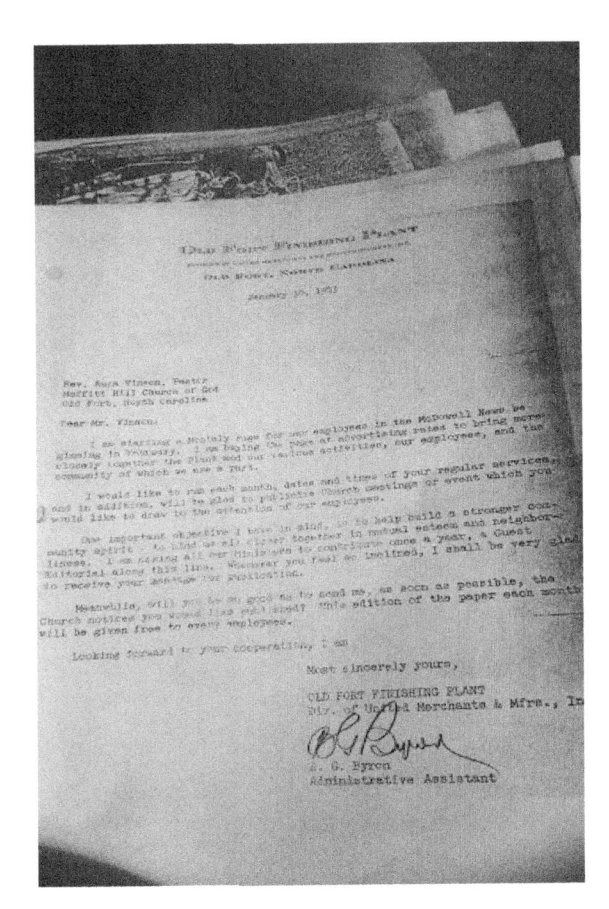

This was the newspaper asking permission to advertise our church in Marion NC.

This picture is very precious to me. That is my daddy, the first man on the left with Albert in front. The very last lady on the end is my mama, holding Aneita. There is Eza and Mauga and grandma Vinson.

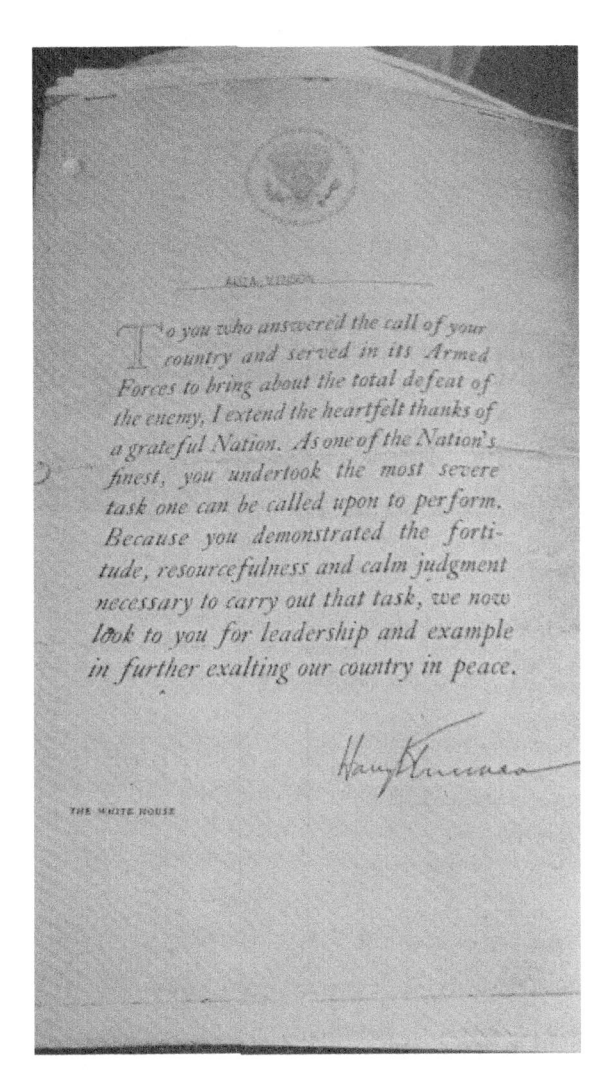

> AUZA, VIDSON
>
> To you who answered the call of your country and served in its Armed Forces to bring about the total defeat of the enemy, I extend the heartfelt thanks of a grateful Nation. As one of the Nation's finest, you undertook the most severe task one can be called upon to perform. Because you demonstrated the fortitude, resourcefulness and calm judgment necessary to carry out that task, we now look to you for leadership and example in further exalting our country in peace.
>
> *Harry Truman*
>
> THE WHITE HOUSE

Auza's Honorable Discharge.

Auza and Albert and Robbie

157

158

Hi, my name is Aneita and I am the fifth child in this story. Mom and dad have now passed away and are missed daily. I felt close to them while I wrote their story. I did add some things that I thought, in between scenes might be nice. This book is mostly fact but the fiction parts come as if I were a fly on the wall, so to speak. Some of the events maybe occurred, maybe not, I was adding some joy to the theme. Being that my mom was so busy most of the time with us kids. I gave my best to make this a very pleasant book. I hope you enjoyed and thank you for reading Creolia x Six.

Made in the USA
Columbia, SC
20 March 2021